IEE CONTROL ENGINEERING SERIES 54

Series Editors: Professor D. P. Atherton
Professor G. W. Irwin

CONTROL ENGINEERING SOLUTIONS

a practical approach

Other volumes in this series:

CONTROL ENGINEERING SOLUTIONS

a practical approach

Edited by

P. Albertos
R. Strietzel
N. Mort

The Institution of Electrical Engineers

Published by: The Institution of Electrical Engineers, London,
United Kingdom

© 1997: The Institution of Electrical Engineers

The Institution of Electrical Engineers,
Michael Faraday House,
Six Hills Way, Stevenage,
Herts. SG1 2AY, United Kingdom

British Library Cataloguing in Publication Data

A CIP catalogue record for this book
is available from the British Library

ISBN 0 85296 829 9

Printed in England by Short Run Press Ltd., Exeter

Contents

Preface

This book collects together in one volume a number of suggested control engineering solutions which are intended to be representative of solutions applicable to a broad class of control problems. It is neither a control theory book nor a handbook of laboratory experiments, but it does include both the basic theory of control and associated practical laboratory set-ups to illustrate the solutions proposed.

Most of the content was developed, presented and discussed within the European TEMPUS Project IMPACT (IMProvements in Automation & Control Technology), led by the Austrian group of Professor Manfred Horvart at the Vienna University of Technology, to whom we are all grateful for the excellent organisational support provided throughout the project. Nevertheless, further refining, improvement and debugging has been necessary to reach this final form. Such an origin makes the book truly international, with contributors from many different countries and diverse teaching environments.

The main purpose of each contribution is to identify an industrial control problem, to discuss different approaches to solve it and to suggest straightforward laboratory rigs to obtain practical knowledge about this (and related) problems. The common structure of the contributions is broadly based on:

- treatment of a well-defined and motivated industrial control problem;
- outline of the control theory involved;
- discussion of alternative approaches to deal with it;
- basic laboratory set-up to reproduce the scenario (sometimes just by simulation);
- experiments suggested to grasp the basic concepts of the problem;
- illustrative results obtained from the proposed set-up, which may be useful as a guideline for local replications; and
- final comments and conclusions followed by an introductory bibliography.

The book should interest a broad audience. Control engineering students will find potential applications for control theory and workable examples of practical control problems. Most of the laboratory set-ups will be very easy to replicate by control engineering teaching staff, enabling practical activity to complement theoretical and exercise class sessions. Last, but not least, applied control engineers faced with real control problems will find guidelines to approach the solution of their own control problems, including discussion of alternative methods and expected results. We hope that the book will complement conventional theoretical and exercise-oriented control textbooks.

The initial idea for the book was first raised with the co-editor Professor Roland Strietzel, organiser of the first round-table discussion session in the framework of EXACT 93, an IMPACT technical meeting held in Dresden in 1993. However, the idea could not become a reality without the effort of Dr. Neil Mort, whose input has included effort in reviewing and improving the style and language of all but one of the chapters which originated from authors whose native language is not English.

We hope we deserve the confidence the Institution of Electrical Engineers has shown in publishing our project which is intended to fill a small part of the gap between theory and practice in the control field.

Pedro Albertos, Valencia, October 1996

Introduction

In a world where reliability, speed of communication, production competitiveness or product quality requirements, amongst many other technological factors, are more and more demanding, the use of control engineering techniques is an interdisciplinary challenge to be applied in different fields of technology.

Control engineering problems, which were historically approached from the two different viewpoints of either tracking or regulation problems, have deservedly received considerable attention and research over the last few decades. A number of theories and techniques have been developed, but they have not always fulfilled the actual need of the users responsible for getting the systems to work properly under control. To understand each one of these approaches well, specific knowledge is needed and special tools are available, which usually require some theoretical background.

In most of the control problems engineers dealt with, these techniques are not exclusive but complementary, each one leading to a partial solution. It is not only the requirements and design constraints that define the most suitable approach; the "best" solution will probably involve a combination of concepts and techniques, generally studied in different frameworks.

There are a number of steps to take before a control engineering solution can be successfully applied to a given problem. From the initial phase of experimentally building up a model of the process, to the final steps of hardware and software implementation and validation of the designed control system, a number of issues appear. Even where we short-cut the whole process and directly choose a standard control solution already applied to a similar problem, the final steps of integrating controlled system components, tuning the parameters, checking the control requirements and validating the control solution must still be performed.

The main purpose of each chapter in the book is to identify an industrial control problem, discuss different approaches to solve it and suggest easy-to-implement laboratory rigs to obtain practical knowledge about this and related problems. Some of the reported problems and techniques discussed are very broad or complex and would require much more space and time to provide a full understanding. Our purpose is less ambitious; we simply try to define the problem, propose some solutions and provide references for further study, if so required. The common structure of the chapters is broadly based on:

- treatment of a well-defined and motivated industrial control problem;
- outline of the control theory involved;
- discussion of alternative approaches to deal with it;
- basic laboratory set-up to reproduce the scenario (sometimes just by simulation);
- experiments suggested to grasp the basic concepts of the problem;
- illustrative results obtained from the proposed set-up, which may be useful as a guideline for local replications; and
- final comments and conclusions followed by an introductory bibliography.

The book is organised into seventeen chapters, a summary of which follows here:

The first chapter, by J. Picó, P. Albertos and M. Martínez, is devoted to the initial problem of experimental modelling of processes. A typical industrial process, such as a neutralisation tank (or a set of them) is considered. Focusing on techniques for parameter estimation of discrete-time single-input single-output (SISO) models, the basic least-squares family of algorithms is presented and some practical issues in carrying out the experiments are highlighted and discussed. A complementary viewpoint on the solution is presented in Chapter 14.

The second chapter summarises techniques for studying most of the basic control approaches by analogue simulation, from linear to non-linear and from SISO to multivariable (MIMO) systems. The authors, W. Badelt and R. Strietzel, present a versatile analogue set-up to simulate any of these systems and to evaluate the effectiveness of theoretically-designed controllers. The use of measuring devices and actuators allows for a better understanding of the real problems found when implementing a controller.

P. Bikfalvi and I. Szabó present a digital counterpart to the previous chapter. Although restricted to linear SISO systems, they provide the tools for practical implementation of different classical digital controllers, from the basic PID to well-known cascade and predictive controllers.

Wind-up in integrators is also a classical problem which occurs in control solutions. The use of digital computers, both to implement the control and (partially or totally) to simulate the process, makes this problem more common and, sometimes, less noticeable. In Chapter 4, B. Šulc analyses integral wind-up in control and system simulation, and illustrates its consequences on a practical problem of the control of pressure in a vessel.

The processes of diving and flying, as examples of typical unstable systems, pose challenging problems for design of a reliable control system. In Chapter 5, Đ. Juricic and J. Kocijan review the design techniques available to deal with the control of unstable systems, from classical PID control to the use of genetic algorithms. These solutions are also compared with those obtained using linear quadratic regulators with observers. As in most of the chapters, the experiments are carried out on an interesting laboratory set-up.

The control of thermal processes, so common in process industries, receives the attention of several contributors. P. Zítek, in Chapter 6, analyses the particular issue of the problem of delays in the control of temperature and heat flow rate. Delays are widespread in any process involving the transportation of materials, energy or information. The influence of delays in degrading control performance and ways to overcome it are illustrated experimentally on a laboratory-scale heating system where the process dimensions have been chosen to reduce the time intervals (lags and delays).

A typical academic control problem is the inverted pendulum, which is covered by P. Frank and N. Kiupel in Chapter 7. In addition to the explicit non-linear model and well-posed control requirements, the system has a number of properties and difficulties which make it suitable to illustrate the design and implementation of industrial control solutions. The authors describe the different steps, from theoretical modelling based on mechanical laws, through linearisation and state-space controller design.

As mentioned previously, one classical control problem is to keep a controlled variable within certain limits in spite of the presence of disturbances. P. Albertos and J. Salt deal with disturbance rejection in Chapter 8, where they discuss sources of disturbances as well as approaches to counteract them. The basic rig involves a laboratory-scale model to control the position of the end effector of a welding machine, where disturbances produced by changes in load, measurement noise or wheel-wear can be observed.

Most industrial processes are MIMO systems, although the classical approach is to use multiple single-loop control systems, which are simpler to design. Multivariable process control is the topic is reported by N. Mort in Chapter 9. He covers the basic issues of this topic, illustrated on a motor-alternator set, from experimental system identification to controller design, mainly oriented to decoupling the control actions. The use of available software packages, such as MATLAB Control Toolboxes, allows easy controller design, including design in the frequency domain.

M. Voicu, C. Lazăr, F. Schönberger, O. Păstravanu and S. Ifrim analyse the advantages of predictive control versus PID control of thermal treatment processes in Chapter 10. This is another alternative to overcome the difficulties in controlling processes involving time delays. In this experimental set-up, it is shown that in applications where the reference signal is well known in advance, the use of feedforward and predictive control can improve the performance obtained by PID controllers.

Pneumatic cylinders are used extensively in industrial applications as control actuators. Nevertheless, the nonlinearities inherent in the process and the complexity in the model have made their analytical study difficult. K. Janiszowski and M. Olszewski, in Chapter 11, present a number of interesting suggestions to develop state-space adaptive control for such nonlinear systems. A laboratory set-up including a pneumatic cylinder is described and both experimental and theoretical ways of modelling it are discussed. PID, state-space, adaptive and fuzzy logic controllers are suggested and practical hints given to counteract the nonlinearities.

Distributed parameter processes, that is, those whose state depends not only on time but also on spatial coordinates, are rather common in process industry. Real thermal systems and, in particular, continuous kilns (where the material circulates inside a multizone kiln) are relevant to many different applications. This is the control problem covered by B. Rohál-Ilkiv, P. Zelinka and R. Richter in Chapter 12 on distributed process control. In this chapter, the authors analyse the different tools to deal with both distributed process control and boundary control. Two alternative laboratory set ups are proposed.

Ideas of artificial intelligence are being used more and more in the context of control methodologies. From very simple applications to the most complex ones where only approximate knowledge of processes and goals are given, the use of fuzzy logic controllers is becoming very popular. P. Frank and N. Kiupel dedicate Chapter 13 to a demonstration of fuzzy control with an inverted pendulum. The authors also consider applications to a steam turbine and aircraft flight.

In computer-controlled systems, the code required to implement the control algorithm is often a minimal part of that of the real-time application needed to handle the process operation information. This information can easily be used for other purposes, in particular, to develop a supervisory level. Based on the information produced in the operation of an adaptive control system, a supervision strategy is proposed by M. Martínez, P. Albertos, J. Picó and F. Morant in Chapter 14. This supervision can be implemented as a set of rules with a structure similar to that of an expert system, although the simplicity of the reasoning does not require any specific software. A pair of coupled tanks serves as the laboratory set-up to illustrate the results. Nevertheless, the authors devote most of the chapter to discussion of which indices and indicators are the more suitable to perform adaptive control supervision.

Another basic objective of this upper level supervisor is the detection of faults, alarms or malfunctions. This is a crucial issue in many control applications due to the danger of long-lasting hazards or general safety reductions on the operating conditions. P. Frank and B. Köppen-Seliger address this issue in Chapter 15. The three coupled tank system is used as a basic process where the model, control and operation are clearly understood, in such a way that diagnosis concepts based on so-called analytical redundancy can be illustrated.

The last two chapters deal with practical implementation issues related to either hardware or software. C. Lazăr, E. Poli, F. Schönberger and S. Ifrim discuss microcomputer-based implementations for DC motor-drive control in Chapter 16. Two control schemes are compared, the classical cascade control of speed and current, and two-loop parallel control. Flow diagrams and the system layout given will help in replicating this set-up in any laboratory.

Finally, A. Braune introduces some of the basic issues in software design for real-time systems in Chapter 17. The leading application is the programming of an educational robot used as a versatile manipulator, but most of the concepts related to the modular design of real-time software will be very valuable for any of the control applications previously discussed.

As you read the book, you will see that, although both the basic theory of control and the practical laboratory set-ups to illustrate the proposed solutions are outlined, this is neither a control theory textbook, nor a handbook of laboratory experiments. It is more a book about the fusion of control theory and practice and we hope it will be a useful book for general reference and consultation.

Contributors

P. Albertos, M. Martinez,
F. Morant, J. Picó and J. Salt
Department of Systems Engineering,
  Computers and Control
Universidad Politecnica de Valencia
PO Box 22012
E-46071 Valencia
Spain

P. Bikfalvi and I. Szabó
Institute of Information Technology
Department of Process Control
University of Miskolc
3515 Miskolc Egyetemuaros
Hungary

P. M. Frank, N. Kiupel
and B. Köppen-Seliger
Department of Measurement and
  Process Control
Gerhard Mercator University
Bismarckstrasse 81 BB
D-47048 Duisburg
Germany

K. Janiszowski and M. Olszewski
Warsaw Technical University
Institute of Industrial Automatic
  Control
ul. Chodkiewicza 8
PL-02 525 Warsaw
Poland

Đ. Juricic and J. Kocijan
University of Ljubljana
Jozef Stefan Institute
Jamova 39
SLO-61111 Ljubljana
Slovenia

N. Mort
Department of Automatic Control
  and Systems Engineering
University of Sheffield
PO Box 600
Mappin Street
Sheffield
S1 4DU
UK

B. Rohál-Ilkiv, P. Zelinka
  and R. Richter
Department of Automatic Control
  and Measurement
Mechanical Engineering Faculty
Slovak Technical University
Nam. Slobody c. 17
812 31 Bratislava
Slovak Republic

M. Voicu, C. Lăzar,
  F. Schönberger, O. Păstravanu,
  E. Poli and S. Ifrim
Department of Industrial Control
  and Industrial Informatics
Gh Asachi Technical University
  of Iasi
Str. Horia 7-9
RO-6600 Iasi
Romania

R. Strietzel, W. Badelt
  and A. Braune
Department of Electrical Engineering
  and Control Theory
Dresden University of Technology
Mommsenstrasse 13
D-01062 Dresden
Germany

P. Zítek and B. Šulc
Department of Automatic Control
Czech Technical University
Technika 4
CZ-166 07 Prague 6
Czech Republic

Chapter 1

Process model identification

J. Picó, P. Albertos and M. Martínez

1.1   Introduction

Process model identification techniques, where the model takes the form of a parametrised discrete-time transfer function, are acquiring increasing relevance as digital control systems become more widespread.

Many control design strategies rely upon a good process model. Within this framework, many parametric identification algorithms are currently available. Most have a common structure and share many strong and weak points. The goal of this chapter is to enable the reader to become acquainted with the practical use of least squares-based estimation algorithms, as a major representative part of this field. A thorough understanding of their general structure, their solutions and the problems that may be encountered should be acquired by anyone intending to apply these algorithms in a practical situation.

1.2   Control problem

Process control and signal processing problems very often involve the identification of an explicit model of the system under consideration. The model identified may then be used to design signal predictors, interference cancellation filters, equalisers, etc., if signal processing problems are considered, or appropriate control laws, if control problems are considered. Within this field, many control design strategies rely upon the identification of an accurate explicit process model to achieve good performance (e.g. explicit self-tuning regulators). Hence, there is a need for a deep study of identification techniques.

Before going further into the identification techniques, some distinctions will be established among the terms *identification*, *modelling* and *estimation*.

By *modelling* one usually refers to the process of setting up a set of equations — a mathematical model — of the system, based on data not necessarily obtained experimentally, but based on physical principles, empirical relations or other *a priori* information [1]. *Estimation* refers to the process of obtaining the parameters of a given model. Models, in this case, usually take the form of stochastic linear differential or difference equations. On the other hand, the term *identification* refers to the process of selecting a parametrised model, from a set of them, and then obtaining its parameters. That is, identification refers to modelling followed by estimation. The modelling stage associated with process identification is often termed 'model structure determination'.

The classical approaches to process identification deal with the analysis of either the process time response to step or impulse inputs or the process frequency-domain response. In the first case, techniques like the Strejc method lead to approximated models, suitable for controller design purposes. The use of stochastic processes or pseudo-random binary signals allows us to obtain the process impulse response by correlation.

Non-parametric models are also obtained by process frequency response analysis, and some techniques will enable the selection of an approximated parametric model. Of course, this approach is suitable if a frequency-domain controller design method is used.

Today, however, the use of on-line computer-based identification schemes provides the tool to obtain continuous time or discrete time parametric models easily. In turn, there are many controller design techniques based on these models.

Ljung [2] formulated a sequence of questions that any user of identification methods must answer:

(i)   Has system identification anything to offer to my problem?
(ii)   How should I design the identification experiment?
(iii)   Within which set of models should I look for a suitable description of the system?
(iv)   What criterion should I use for selecting that model in the model set that best describes the data?
(v)   Is the model obtained good enough for my problem?

Answering these questions is not an easy task; the answers depend on the knowledge that one has beforehand about the system to be identified, on the use foreseen for the model identified, and on the powers and limitations of the available techniques[1].

In this chapter some practical aspects concerning process identification will be reviewed, and tested on a waste-water pilot plant.

In sulphuric acid production plants, one of the main concerns is the treatment of the waste-water, which leaves the production process with a high temperature and low pH value. Control actions have to be taken so as to reduce the temperature and drive the pH to values within an interval ranging from pH 6.5 to pH 8.5. In Figure 1.1, a simplified scheme of a pilot plant representing this process is depicted.

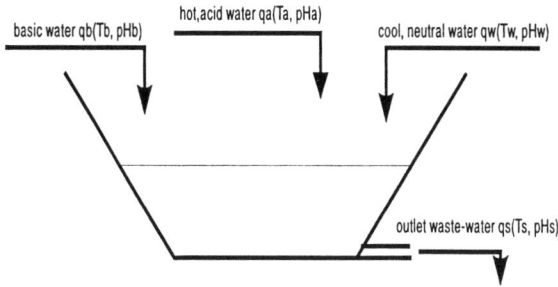

Figure 1.1   Pilot plant scheme

The warm acidic waste-water is introduced into a mixing tank to be cooled and neutralised. For that purpose, a basic inlet flow and a neutral cool water flow are added. The process is highly nonlinear and the variables of interest are coupled.

The classical control procedure operates through actuation upon the cool water inlet flow (qw) to control the outlet flow temperature (Ts). On the other hand, by modifying the basic water inlet (qb), the outlet pH (pHs) may be controlled [3].

Faced with the problem of designing appropriate controllers, a good plant model is one of the first requirements. More specifically, in the case we are concerned with, identification of the relationships between the variables Ts/qw and pHs/qb is required. For this purpose, modelling based on the knowledge of the physical and chemical behaviour laws of the process may be undertaken. (The basic equations for the process are described in Appendix 1.A.) Nevertheless, the identification of discrete-time parametric models through a computer offers the following advantages:

- Modelling from physical and chemical laws, as proposed before, besides needing a good knowledge of the process, will eventually require knowledge of physical and chemical process constants which may be difficult (if not impossible) to obtain;
- Parametric identification methods allow the modelling of process disturbances;
- Low-magnitude excitation signals may be used, which is important in general (and even more so in the case of nonlinear processes, such as the one considered here);
- Non-parametric models of the step or frequency response type may be obtained accurately from the parametric models [4];
- Recursive on-line identification algorithms may be used if required (i.e. in adaptive control).

Henceforth, parametric identification algorithms will be used to identify the aforementioned relationships between Ts/qw and pHs/qb.

1.3   Technical approaches

Parametric identification methods, either recursive or not, have been widely studied for many years [2,4–7]. All the methods can be grouped into two categories [4]:

(1) identification methods based on the whitening of the prediction error;
(2) identification methods based on the decorrelation of the prediction error and the regression or observation vector.

Within each category, different methods exist which, in turn, can face different model structures. Apart from SISO (single-input, single-output) structures, multivariable versions of the algorithms are available to estimate MIMO (multi-input, multi-output) structures [8]. They will not be considered in this chapter. Attention will be restricted to two of the most widely used methods, the least squares (LS) and the extended least squares (ELS), both based on the whitening of the prediction error.

The least squares method can be used to tackle models with a structure given by

$$A(q^{-1})y(k) = B(q^{-1})q^{-d}u(k) + v(k) \qquad (1.1)$$

where $y(k)$, $u(k)$ and $v(k)$ are the input, output, and disturbance at instant k, and A and B are polynomials defined as

$$A(q^{-1}) = 1 + a_1 q^{-1} + a_2 q^{-2} + \ldots + a_n q^{-n}$$
$$B(q^{-1}) = b_1 q^{-1} + b_2 q^{-2} + \ldots + b_m q^{-m}$$

where q^{-1} is the backward shift operator.

The disturbances are described by the stochastic process

$$v(k) = H(q^{-1})e(k) \qquad (1.2)$$

where $H(q^{-1})$ is a rational function which in this case equals the identity (i.e., $H(q^{-1}) = 1$) and $\{e(k)\}$ is white noise statistically defined by

$$E[e(k)] = 0$$
$$E[e(k)e(k+T)] = \sigma^2 \delta(T)$$

The extended least squares algorithm attempts to estimate the parameters of a model structure where

$$H(q^{-1}) = C(q^{-1})$$
$$C(q^{-1}) = 1 + c_1 q^{-1} + c_2 q^{-2} + ... + c_{nc} q^{-nc}$$

(1.3)

As for the model structure used by the LS algorithm, the process output can be expressed as

$$y(k) = \alpha^T(k)\theta(k) + e(k)$$

(1.4)

where the regression or observation vector and the parameter vector are defined respectively by:

$$\alpha^T(k) = [-y(k-1),....,-y(k-n), u(k-d-1),..., u(k-d-m)]$$
$$\theta^T = [a_1,...,a_n, b_1,...b_m]$$

It is well known that the least squares algorithm provides the estimated parameters θ at instant k so that the criterion

$$L(\theta,k) = \sum_{i=1}^{k} [y(i) - \alpha^T(i)\theta(i)]^2$$

(1.5)

is minimised. The estimated parameter vector is given by [2,7,9]

$$\theta(k) = [\psi^T(k)\psi(k)]^{-1} \psi^T(k) Y(k)$$

(1.6)

where

$$\psi(k) = [\alpha(1),\alpha(2),...,\alpha(k)]^T$$
$$Y(k) = [y(1), y(2),..., y(k)]^T$$

The recursive version of this algorithm can be readily obtained [4,9], yielding the recursive least squares (RLS) algorithm:

$$\theta(t+1) = \theta(t) + K(t+1)[y(t+1) - \alpha(t+1)^T \theta(t)]$$

(1.7)

$$K(t+1) = \frac{P(t+1)\alpha(t+1)}{1 + \alpha(t+1)^T P(t)\alpha(t+1)}$$

(1.8)

$$P(t+1) = P(t) - \frac{P(t)\alpha(t+1)\alpha(t+1)^{T} P(t)}{1 + \alpha(t+1)^{T} P(t)\alpha(t+1)} \qquad (1.9)$$

where P is the so-called inverse covariance matrix.

The recursive extended least squares (RELS) algorithm follows the same structure given by Equations 1.7, 1.8 and 1.9, the only difference being the regression vector and estimated parameter vector components, which must be extended in order to estimate the coefficients of polynomia $C(z^{-1})$ Thus, the new vectors become:

$$\alpha^{T}(k) = \left[-y(k-1), \ldots, -y(k-n), u(k-d-1), \ldots, u(k-d-m), e_{p}(k-1), \ldots, e_{p}(k-n_{c}) \right]$$

$$\theta = \left[a_{1}, \ldots, a_{n}, b_{1}, \ldots b_{m}, c_{1}, \ldots, c_{n_{c}} \right]$$

where the prediction error

$$e_{p}(k+1) = y(t+1) - \alpha(t+1)^{T}\theta(t) \qquad (1.10)$$

is used as an estimate of the white noise signal e.

For both algorithms, a set of convergence conditions must be met [2,7].

Other varieties of estimation algorithm, apart from the least squares, may be used. The 'approximate maximum likelihood' and the 'instrumental variable' algorithms are widely used. The first of these performs better than least-squares identification in the case of coloured noise [10].

The instrumental variable algorithm attempts to decorrelate the residuals from the regression vector, thus avoiding an estimation bias. This is achieved by considering, at the regression vector, not the process output but an auxiliary variable highly correlated with the undisturbed process output. In principle, there is no guarantee that all choices of instrumental variables will provide good identification properties.

1.4   Discussion and laboratory experience

In the following section, knowledge from practical laboratory experience with the pilot plant will enable a thorough discussion of classical topics related to the application of identification algorithms to practical situations. First, the laboratory set-up will be briefly sketched.

1.4.1 Laboratory set-up

The pilot plant used for the laboratory experiments was built using very cheap and basic materials as indicated below.

Three tanks were used. Two small tanks are used to mix acid (H_2SO_4 40% dissolved in water) and bleach with water, to obtain homogeneous inlet flows. A third, larger tank, which may have variable sections, is used as a general mixing tank. Pneumatic valves are used to manipulate the flows. The liquid level is measured using an electronic differential pressure sensor. All sensors used give 4–20 mA output current, which is fed to current-to-pneumatic converters (3–15 psi). A PC with a PCL-812PG PC-LabCard completes the equipment.

Two different operating points of the pilot plant, given by the following values, were used:

	Basic (qb)	Acid (qa)	Neutral (qw)
Flow ($cm^3 s^{-1}$)	20.0 (27.3)	11.11 (25.0)	63.0 (70.0)
Temp. (°C)	18.0	50.0	17.0
pH	10.0	4.0	7.5

Instead of acidic water, another interesting possibility for similar work would consist of using chemical salts and controlling the liquid's conductivity.

1.4.2 Sampling period selection

Parameter estimation procedures involve the pre-selection of certain algorithm parameters. For discrete-time models, the first issue to tackle is the sampling period selection. This will depend either on the time constants of the process (if the aim is just to model it) or on the desired time constants of the controlled closed-loop system if the main goal is to model it in order to design an appropriate controller.

As is well known, the shorter the sampling period t_s, the smaller the magnitude of the numerical values obtained for the coefficients of the polynomial $B(z^{-1})$. The coefficients of $A(z^{-1})$, on the other hand, will grow with its roots tending towards unity.

Figure 1.2 shows the evolution with time of the different process variables for variations in the neutral water and basic inlet flows. Over-damped behaviour can be observed for both *Ts* and *pHs*, with apparent time constants of the order of 500 s for *Ts* and 250 s for *pHs*. Therefore, a first choice of $t_s = 25$ s seems adequate for both loops. Later, once the model structure has been chosen correctly, identification may be carried out for different sampling periods, therein selecting the most appropriate value. Practical issues, such as available hardware and software restrictions, desired dynamics for the controlled process, numerical conditioning of the estimated polynomial coefficients etc., will then be taken into account.

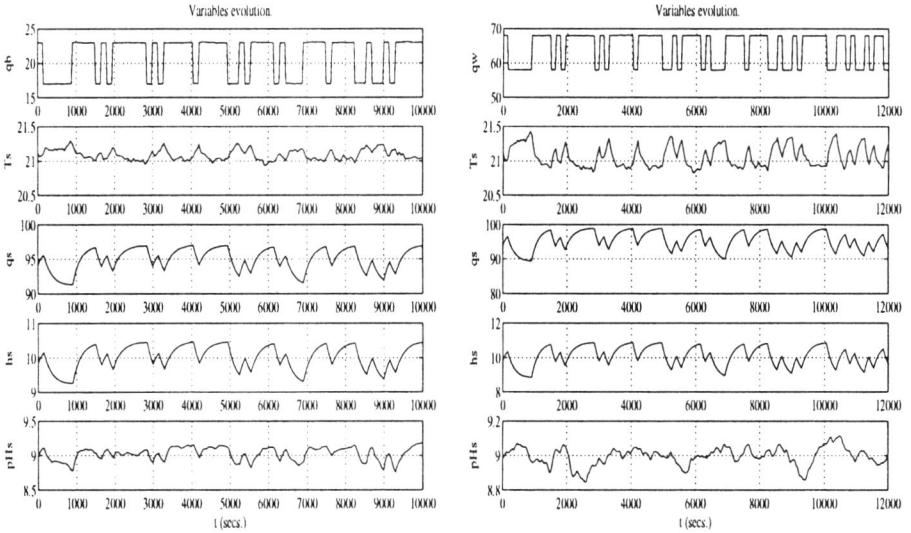

*Figure 1.2   Evolution of plant outputs for variations in the basic inlet flow
(qb, left) and neutral inlet flow (qw, right)*

1.4.3   Data conditioning

Real data obtained from the process must be properly conditioned, prior to its application in the estimation algorithms. Three main points can be considered in this respect.

1.4.3.1   High frequency signal filtering
The first step involves filtering out high frequency signal components so as to avoid sampling aliases [4,11]. Additional filters may be used if enhancement of a specific data frequency range is desired [12].

1.4.3.2   Offset and drift removal
The parametric structures to be identified correspond to dynamic models, expressing the variation of process variables around their operating point. Therefore, offset levels (which might not necessarily correspond to the operating point) and non-stationary DC components (drift) must be removed from the data.

Two different situations may appear. If off-line estimation is considered, data are available at once, thus allowing computation based on the whole data set. In the case of on-line estimation, input-output data are available for each sampling period, and recursive calculations are required. The following approaches are possible:

•   Off-line estimation

—   Offset levels can be eliminated by subtracting the mean signal level;

—   Drifts can be eliminated using filtered input-output variations [4]:

$$y_f(t) = \frac{y(t) - y(t-1)}{1 + f_1 z^{-1}}, \quad u_f(t) = \frac{u(t) - u(t-1)}{1 + f_1 z^{-1}} \qquad (1.11)$$

with $-0.5 \le f_1 \le 0$.
This method can also be used in on-line identification.

•   On-line estimation:

In addition to the previous method for drift removal, the following, which can remove both stationary and non-stationary DC components, are also available:

(1)   use of data differences between samplings:

$$y_f(t) = y(t) - y(t-1), \quad u_f(t) = u(t) - u(t-1) \qquad (1.12)$$

(2)   calculation of the mean level approximately through a recursive first-order filter:

$$\bar{y}(t) = (1 - \beta)\bar{y}(t-1) + \beta y(t)$$
$$y_f(t) = y(t) - \bar{y}(t)$$

(3)   use of identification algorithms modified so as to estimate the DC components [5].

In Figure 1.3, the offset removal for the temperature loop is shown.

1.4.3.3   Scaling of inputs and outputs
If the levels of inputs and outputs are very different from each other, then the convergence speed of the estimated parameters associated with the model polynomials will be very different. One solution to this problem consists of scaling data by multiplying inputs or outputs by the appropriate constant. Obviously the stationary gain identified will not be correct, and the estimated parameters of $B(z^{-1})$ will have to be re-scaled accordingly.

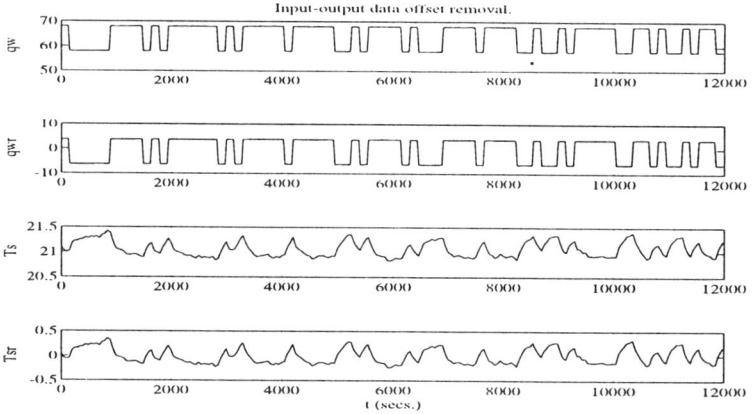

Figure 1.3   Offset removal for Ts and qw

1.4.4   Model structure selection

A correct model structure has to be determined for each of the process loops to be identified. In Section 1.3, two basic model structures were considered:

$$A(z^{-1})y(k) = B(z^{-1})z^{-d}u(k) + e(k) \qquad (1.13)$$

$$A(z^{-1})y(k) = B(z^{-1})z^{-d}u(k) + C(z^{-1})e(k) \qquad (1.14)$$

So, determining the model structure involves:

(1)   determining the disturbance structure; and
(2)   determining the process structure, i.e. the correct orders for polynomials A and B and the correct delay d.

A set of tests are available for evaluating these points [4,7,12], including:

Determination of	Test
A and B orders	Loss Function
	Determinants ratio
	Residuals variance decrease
Delay, d	Loss function
	Poles-zeroes cancellation
Perturbations structure	Residuals analysis

Let us consider the pilot plant *pHs/qb* and *Ts/qw* loops and apply two of the most widely used tests for the determination of the polynomial order; the loss function and the residuals variance decrease methods. For delay determination, the loss function test will be applied. Subsequently, analysis of the algorithm residuals will be carried out to decide on the disturbance structure.

First, a model structure given by Equation 1.13 will be considered. A set of LS algorithms corresponding to different values of n_a, n_b and d is selected, and the sum of the squared prediction errors — i.e. the loss function — is computed over a data set for each of the runs. The structure giving the least value of the loss function, while maintaining simplicity, is chosen.

It should be taken into account that if a model structure is evaluated on the same data set on which it was estimated, the loss function will always decrease as the model structure orders (n_a and n_b) are increased (whereas this is not true for the delay d). However, from a given value of the number of parameters $n_a + n_b$ upwards, the decrease in the loss function becomes small, indicating that an increase in the number of parameters does not improve performance.

Several tests, such as the Akaike's final prediction error, the information theoretic criterion [12] or the residuals variance decrease test, can be used to make a decision.

Some practical guidelines to determine the required values, if there is no *a priori* knowledge about the process, are:

- n_a: typical values span the interval $1 \leq n_a \leq 3$

- n_b: to tackle fractional delays $n_b \geq 2$

In Figure 1.4 the results for the *pHs/qb* loop are shown when the loss function test is used. Observe that the results given by the test are incremental, in the sense that once an optimal value has been reached for one of the parameters (n_a, n_b or d), the test can be carried out over the other parameters consecutively, hence reaching the global optimum by consecutively reaching the optimum for each problem coordinates.

Some results are shown for the residuals variance test used on the *pHs/qb* loop in Figure 1.5 (left). A value $d=0$ was assumed and different orders n_a and n_b were tested, corresponding the best fit to $n_a + n_b = 4$. Should the elbow at the optimum be not as neat as the one obtained in this case (e.g. *Ts/qw* loop, as shown in Figure 1.5 (right)), criteria based on information gain and structure simplicity may be used [4].

The second step is to deal with the disturbance model selection. As mentioned in Section 1.3, the LS algorithm aims at whitening the residuals (prediction error) sequence. Therefore, if the correlation function of residuals is calculated, it should vanish except for the autocorrelation. In practice, values of the correlation function will not vanish for time shifts greater than one, but should be negligible. Statistical criteria may be used to evaluate the correlation functions. Thus, in [4] the residual prediction error normalised covariance is calculated as:

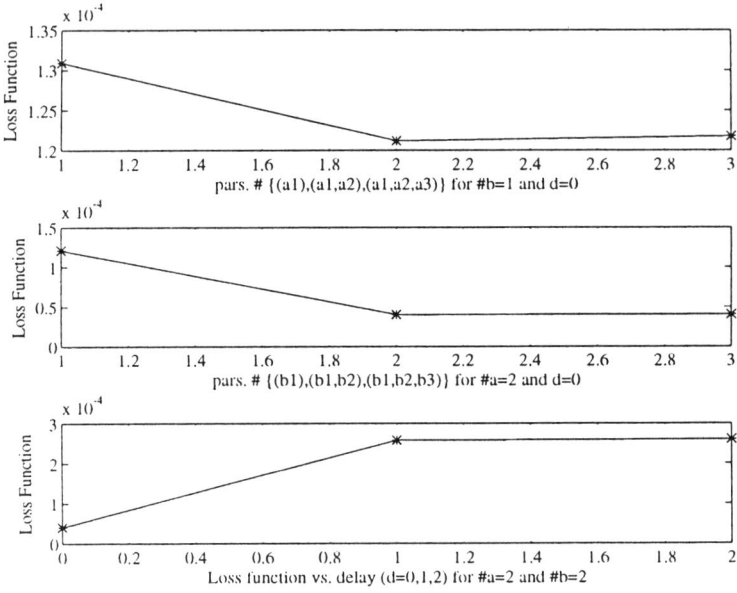

Figure 1.4   Loss function test results for the pHs/qb loop

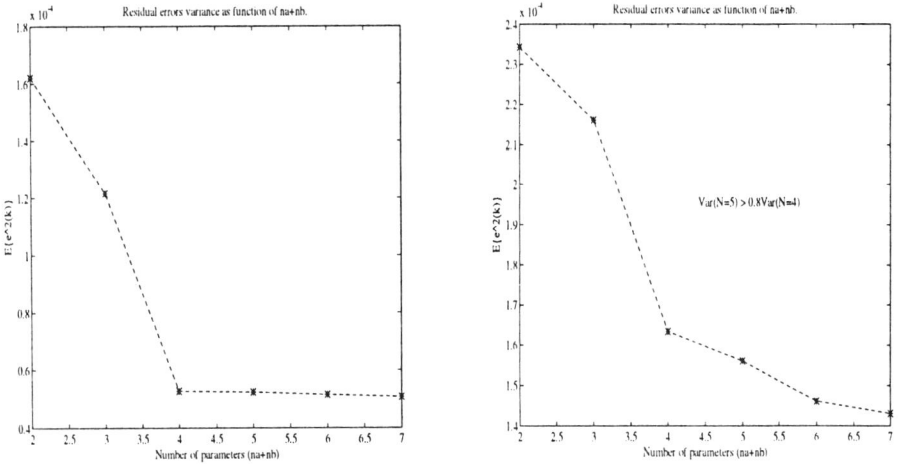

Figure 1.5   Residuals variance test results for the pHs/qb (left) and Ts/qw (right)

$$NC_v(i) = \frac{C_v(i)}{C_v(0)}, \quad i = 1,\ldots,n_a,\ldots$$

If $NC_v(i)$ obeys the Gaussian distribution $N(0,1/\sqrt{N})$, where N is the data set length, then the relationship

$$|NC_v(i)| = \frac{2.17}{\sqrt{N}} \tag{1.15}$$

should hold for a confidence interval of 3% (other confidence intervals may be considered).

The magnitude of the residual prediction errors must not be very low compared to the process output magnitude for this test to be significant [4]. Thus, a relation between the typical deviations of the conditioned process output and those of the residuals, given by $\sigma(output)/\sigma(residuals) < 60$ dB, should hold. If the converse situation holds, then the bias in an LS due to consideration of a simplified disturbance model can be neglected.

The results of the residuals correlation test for the two pilot plant loops considered are shown in Figure 1.6. An improvement is obtained in both loops if an extended model is identified. The result is clear for the *Ts/qw* loop (right). For the *pHs* loop (left) a simple LS-like model (Equation 1.13) may be considered.

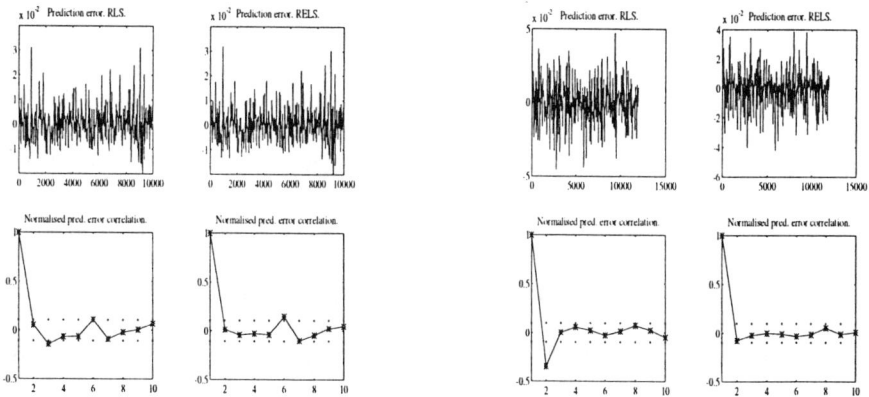

Figure 1.6   *Residuals and their correlations for the pHs/qb (left) and Ts/qw (right)*

1.4.5   Signal excitability

For a dynamic model to be correctly estimated, input signals to the process must be sufficiently exciting [2,12]. More precisely, all process dynamic modes in the frequency range of interest must be persistently excited. Hence, process input signals should have a frequency content sufficiently rich to span a range wide enough to correctly identify both the fast dynamic modes and the stationary gain. Pseudo-random binary sequence signals (PRBS) are widely used for this purpose.

The estimated parameters for *Ts/qw* are shown in Figure 1.7.

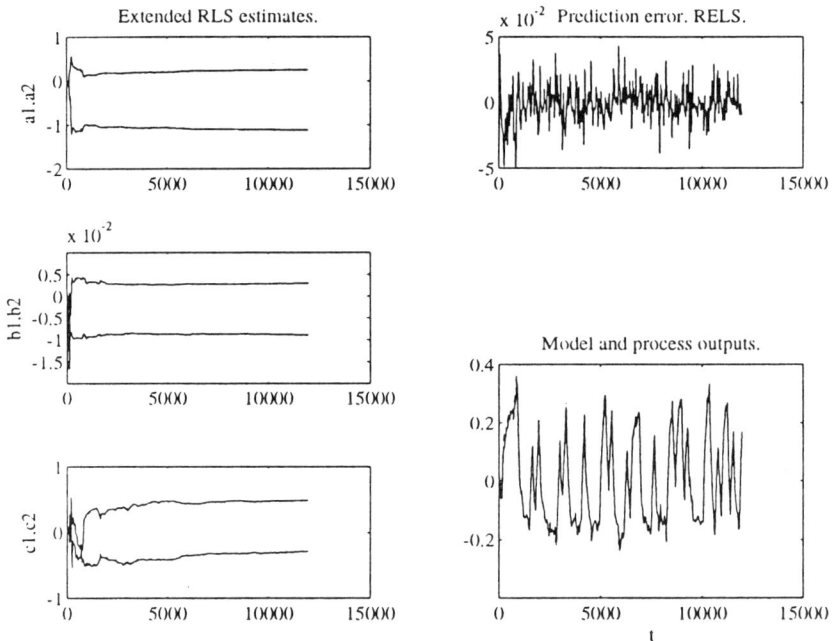

Figure 1.7   Parameter estimation for Ts/qw. Recursive extended LS algorithm

If the process input is not sufficiently exciting, numerical problems will arise in the inverse covariance matrix (Equation 1.9), leading to poor estimation or even to the so-called 'parameter blow-up' if a forgetting factor is being used (see Section 1.4.6). Possible solutions, if one has to cope with a process where there is lack of signal excitability, include the use of algorithms wherein *a priori* known values of the parameters are weighted in the loss function [13,14].

1.4.6   *Parameter tracking and forgetting factors*

An important issue when the estimated model has to be used on-line, and needs to be updated when possible, is that of parameter tracking. Process parameter changes are not unusual and three main causes can be considered:

(1) drifts in the process dynamic behaviour caused by temperature, time etc.;
(2) changes in the process physical configuration, leading to changes in its dynamic behaviour (e.g. opening or closing of valves in liquid tanks). The change in the dynamic behaviour may be rather fast in this case;
(3) large reference excursions in nonlinear systems. In this case the linearised transfer functions around the new operating point may differ significantly from the previous ones.

1.4.6.1   *Algorithm fading*

Use of a recursive identification algorithm is insufficient to ensure tracking of process parameter changes. In fact, if the loss function given by Equation 1.5 is analysed, it becomes clear that, as time passes and therefore the quantity of available data increases, the relative weight of the new incoming data, with respect to the whole data set on which Equation 1.5 is minimised, becomes negligible. Thus, as time passes, the algorithm becomes insensitive to new data and so to possible process dynamic changes.

This effect is known as algorithm fading and can also be explained from an analysis of the inverse covariance matrix Equation 1.9. The second term on the right side of the equation is a quadratic term. Hence, if process input signals are sufficiently exciting, the P matrix components will decrease with time. The more excited the process, the quicker the decrease of P values. From Equations 1.7 and 1.8 it is clear that this will have a direct influence on the values of the estimated parameters. To monitor algorithm fading, the trace of the inverse covariance matrix can be used, since it gives an indirect measure of the magnitude of its components.

In Figure 1.8 the pilot plant is driven for 7000 s around an operating point where nonlinearities are extreme. Subsequently, the basic inlet flow is increased so that the plant works around an operating point with a rather different dynamic behaviour. As can be seen in Figure 1.9, this change in dynamic behaviour is not reflected in a corresponding change in the estimated parameters, as it should be. The cause of this can be found by inspecting the covariance matrix trace, which shows that the algorithm has faded.

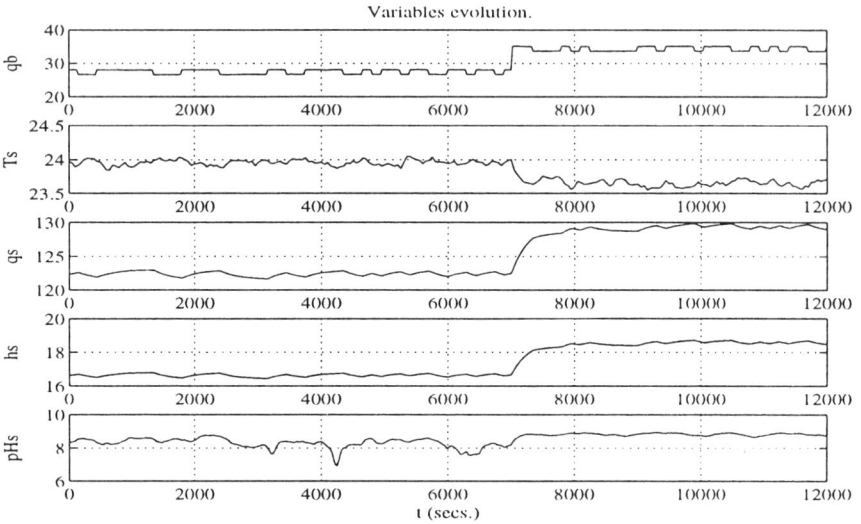

Figure 1.8   Evolution of variables for the pHs/qb loop. The operating point is changed at t = 7000 s

From the previous comments about algorithm fading and its relationship with the loss function, it turns out that to avoid the first, new data should somehow be weighted more than old data. This can be achieved with the use of forgetting factors.

1.4.6.2   Forgetting factors
Forgetting factors are simply a way of weighting data in the loss function to be minimised by the estimator. They can take several forms [2,14,15] including fixed and variable, directional etc. The most widely used type, due to its simplicity, is the so-called exponential window, which uses the loss function:

$$L(\theta,k) = \sum_{i=1}^{k} \lambda^{k-i} \left[y(i) - \alpha^{T}(i)\theta(i) \right]^{2} \qquad (1.16)$$

where λ is the forgetting factor. From this cost function the following recursive equations are obtained:

$$\theta(t+1) = \theta(t) + K(t+1)\left[y(t+1) - \alpha(t+1)^{T}\theta(t) \right] \qquad (1.17)$$

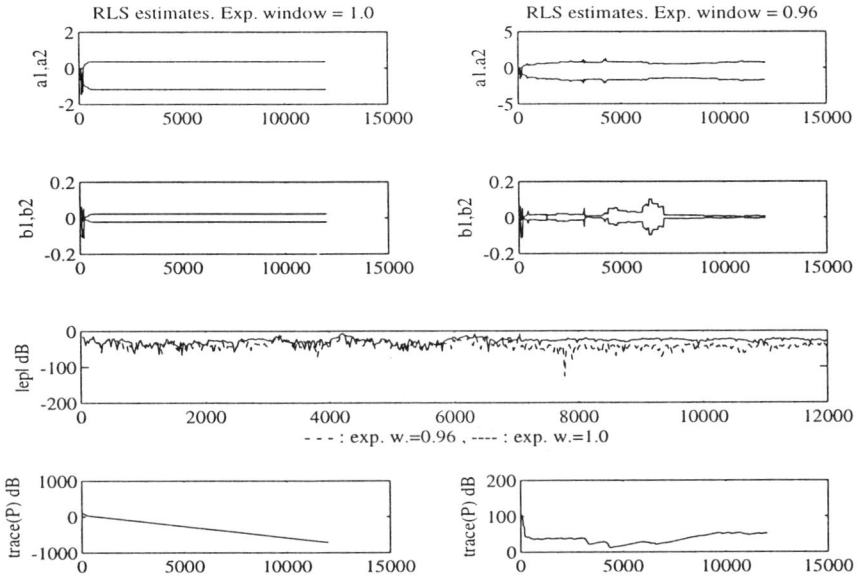

Figure 1.9   *Parameter estimation for pHs/qb; RLS algorithms with and without exponential window*

$$K(t+1) = \frac{P(t+1)\alpha(t+1)}{\lambda + \alpha(t+1)^T P(t)\alpha(t+1)} \qquad (1.18)$$

$$P(t+1) = \frac{1}{\lambda}\left[P(t) - \frac{P(t)\alpha(t+1)\alpha(t+1)^T P(t)}{\lambda + \alpha(t+1)^T P(t)\alpha(t+1)} \right] \qquad (1.19)$$

The results obtained from the pilot plant, when an exponential window is used, are shown in Figure 1.9. During the first 7000 s, the estimated parameters suffer rather large variations at the time instants where the input changes, indicating a strongly nonlinear behaviour around the corresponding operating point. Smoother parameter changes could be obtained by increasing the exponential window λ. When the plant is driven to the second operating point, only the algorithm using a forgetting factor is capable of tracking the process parameters. Thus, if the module of the algorithm residuals in decibels is plotted for both algorithms, it can be observed that its value is appreciably higher if no exponential window is used.

The use of forgetting factors may introduce problems. If the process inputs are not sufficiently exciting, then the quadratic term in Equation 1.19 becomes small, and thus

$$P(k+1) \geq P(k)$$

In fact, it may become negligible, so that

$$P(k+1) \approx \frac{P(k)}{\lambda} \qquad (1.20)$$

If λ is small the gain term in Equation 1.17 will eventually have large fluctuations, leading to similar effects in the estimated parameters, which is known as 'blow-up'.

Additionally, if the forgetting factor is too low, noise tracking by the estimated parameters will appear as a side-effect. Therefore, relatively high values are used (e.g. 0.96–0.98 for exponential windows), although they may be temporarily lowered so as to improve process parameter tracking. In Figure 1.10 it can be seen how a value that is too low for the exponential window λ causes very noisy estimated parameters, although in this particular case the variation associated with the process input is high enough to prevent the estimated parameters from blowing-up. Some of these problems will also be considered in chapter 14.

1.4.7 Closed loop identification

Two main situations may be considered when identifying a process in a closed loop structure [7]:

(1) Indirect process identification — a model of the closed loop is identified. Afterwards, assuming that the controller is known, the process model is calculated by deconvolution;
(2) Direct process identification — the process model is directly identified. The controller need not be known.

Within direct process identification, two main alternatives can also be considered if the process input and output are measured for identification:

(1) no external disturbance is applied;
(2) an external disturbance (measurable or not) is applied.

In the first case a set of identifiability conditions must be met to ensure convergence. For the second, if the additional external disturbance added to the process input is sufficiently exciting and uncorrelated with the process noise e, only the orders of the process must be known, and no additional identifiability conditions are required.

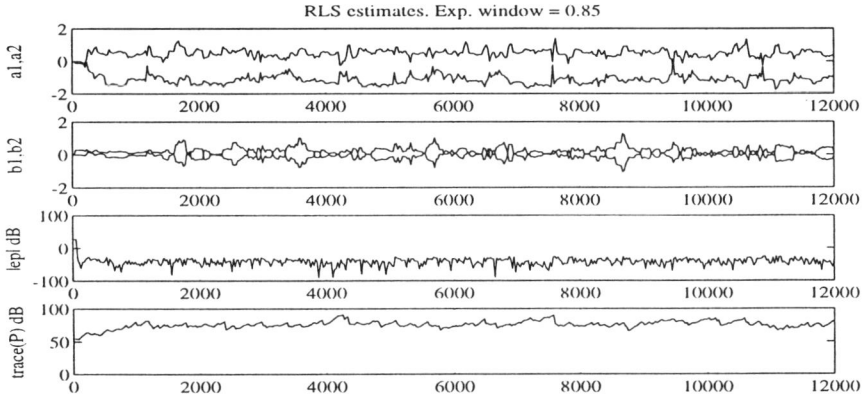

Figure 1.10   Parameter estimation for Ts/qw; RLS algorithms with low exponential window

Recently, it has been shown that the best model for control design cannot be derived from open-loop experiments alone [16]. If the goal of process identification is to use it to implement a controller, the controller to be implemented should be taken into account by the process identification algorithm. However, in most cases, this controller is not available. Hence, iterative process identification and controller design methodologies have recently been devised [17,18].

1.5   Conclusions

Some practical issues concerning process model parameter estimation have been reviewed and applied to a practical case. The first conclusion that may be drawn is the rather iterative character of the estimation process. That is, it involves the selection of a set of choices (model structure, kind of algorithm etc.) beforehand, which must be tested and adjusted. A high degree of precision in modelling the process can be reached at the cost of increased model complexity. Thus, a trade-off must be established based on the final aim of the identified model.

On the other hand, to cope with complex, variable dynamics, the estimator tuning parameters (forgetting factors etc.), the model structure and even the estimating algorithm employed must be adapted on-line. Therefore, a supervisory level is often required for the correct functioning of the overall system (this is especially required when using estimators in a framework of adaptive control).

1.6   References

1   BOHLIN, T.: 'System identification: prospects and pitfalls', (Springer-Verlag, Berlin, 1991)

2   LJUNG, L. and SODERSTRÖM, T.: 'Theory and practice of recursive identification', (MIT Press, 1983)

3   MARTÍNEZ, M., *et al.*: 'Waste-water treatment test plant control: computer aided modern control teaching', *IFAC Trends in Control and Measurement Education*, 1988, Swansea, UK, pp. 53–59

4   LANDAU, I. D.: 'System identification and control design', (Prentice-Hall, 1990)

5   ISERMANN, R.: 'Parameter adaptive control algorithms: a tutorial', *Automatica*, 1982, **18**, pp. 513–528

6   ISERMANN, R., and LACHMANN, R.: 'Parameter-adaptive control with configuration aids and supervision function', *Automatica*, 1985, **21**, (6)

7   ISERMANN, R.: 'Digital control systems, 2nd edn.', (Springer-Verlag, Berlin, 1991)

8   GOODWIN, G. C. and SIN, K. S.: 'Adaptive filtering, prediction and control', (Prentice-Hall, 1984)

9   WELLSTEAD, P. E., and ZARROP, M. B.: 'Self-tuning systems, control and signal processing', (Wiley & Sons, 1991)

10   JOHANSSON, R.: 'System modeling and identification', (Prentice-Hall, 1993)

11   ÅSTRÖM, K. J. and WITTENMARK, B.: 'Computer controlled systems: theory and design', (Prentice-Hall, 1988)

12   LJUNG, L.: 'System identification: theory for the user', (Prentice-Hall, 1987)

13 MORANT, F. and ALBERTOS, P.: 'An algorithm for parameter estimation with multiple undefined solutions: the blending problem'. 7th *IFAC Symp. on Identification and System Parameter Estimation*, York, UK, 1985

14 LAMBERT, E. P.: 'Process control applications of long-range prediction'. *Report OUEL 1715/87*, 1987, Dept. Eng. Science, Oxford University, UK

15 FORTESCUE, T. L. *et al.*: 'Implementation of self tuning regulator with variable forgetting factors', *Automatica,* 1981, **17**

16 SCHRAMA, R.: 'Accurate identification for control: the necessity of an iterative scheme', *IEEE Trans. on Automatic Control*, 1991, **37**, (7), pp. 991–994

17 ALBERTOS, P. and PICÓ, J.: 'Iterative controller design by frequency scale experimental decomposition', *Proc. of the 32nd Conf. on Decision and Control*, San Antonio, Texas, 1993

18 LEE, W., ANDERSON, B., KOSUT, R. and MAREELS, I.: 'On robust performance improvement though the windsurfer approach to adaptive robust control', *Proc. of the 32nd Conf. on Decision and Control*, San Antonio, Texas, 1993

1.7 Appendix: Process physico-chemical behaviour laws

The physical system may be described by the following equations:

• Flow balance:

$$q_b + q_w + q_a - q_s = \frac{d(Ah_s)}{dt}$$ (1.21)

where

q_a : acidic waste-water inlet flow
q_w : neutral water inlet flow
q_b : basic water inlet flow
q_s : waste-water outlet flow
h_s : mixing tank liquid level
A : mixing tank section at the liquid level

• Mixing tank outlet flow:

$$q_s = K\sqrt{h_s} \tag{1.22}$$

$$K = K_p s\sqrt{2g} \tag{1.23}$$

where

K_p : discharge coefficient at the mixing tank output
s : output section
g : gravity constant.

- Thermal balance:

$$C_a T_a q_a + C_b T_b q_b + C_w T_w q_w = C_s T_s q_s + C_s \frac{d(MT_s)}{dt} + K'e(T_s - T_0) - C_r \tag{1.24}$$

where

C_a, C_b, C_w, C_s : specific heats of the corresponding flows
T_a, T_b, T_w, T_s : temperatures of the corresponding flows
C_r : heat of reaction
M : total mass in the mixing tank
K' : heat conduction transmission coefficient
e : mixing tank surface thickness
T_0 : environmental temperature

An approximation can be obtained by assuming all specific heats are equal to 1 cal/g°C and that the heat of reaction is negligible.

- Ionic balance:

$$q_a[OH^-]_a + q_b[OH^-]_b + q_w[OH^-]_w =$$
$$q_s[OH^-]_s + q_a[H^+]_a + \frac{d(Ah_s[OH^-]_s)}{dt} \tag{1.25}$$

where

pH_s, pH_a, pH_b, pH_w : pH of the corresponding flows.
$[OH^-]_a, [OH^-]_b, [OH^-]_w, [OH^-]_s$: oxidril concentration
$[H^+]_a$: concentration of H^+ ions at the acid inlet flow

If all the species are dissociated, then:

$$q_b 10^{pH_b - 14} + q_w 10^{pH_w - 14} = q_s 10^{pH_s - 14} + q_a 10^{-pH_a} + \frac{d(Ah_s 10^{pH_s - 14})}{dt} \tag{1.26}$$

Chapter 2

Analogue controller design

W. Badelt and R. Strietzel

2.1 Introduction

A high percentage of successful control engineering solutions are implemented by conventional controllers, such as single-loop, multi-loop, single-input single-output (SISO), multiple-input multiple-output (MIMO) and multi-step controllers. In many cases the PID structure is used.

Experiments are important in the process of developing relevant knowledge, experience and skills. These experiments have to demonstrate and develop the connections between theory and practice transparently.

A special electronic analogue computer facilitates a great number of instructive experiments on single-loop, multi-loop, multi-variable and multi-step control systems. At different points in the system, signals are measured and their concordance with theoretical results can be assessed. In contrast with digital computer techniques, simulation using analogue systems and signals is often more closely related to real-world continuous processes. Experience in measurement is also obtained and characteristic system behaviour can be recognised.

The usefulness of this type of simulation of the transfer behaviour of systems and components is shown with reference to different examples.

2.2 Motivation

Applying their knowledge of system and control theory the students experimentally grasp the behaviour of the different transfer elements and connections between them. The experiments help to understand:

— the properties of transfer elements;
— the stationary and dynamic behaviour of different control structures;

— the command control and the influence of the reference input;
— the suppression of disturbances;
— the stability of systems.

Using given plant models, different types of analogue controllers can be used and the properties of the control loop can be studied. Tuning rules using transient responses are applied and the resulting controllers are tested by observation of these responses.

Experiments using multi-loop and multi-variable systems are also possible and the advantages and disadvantages of multi-loop systems can be recognised. Use of a two-variable system allows the influence of the coupling coefficient to be studied applying different design methods.

Besides a sound understanding of elementary control structures the student gains experience in experimentation and measurement.

2.3   Technical approaches

In the following the properties of different types of controllers are presented as a basis for practical experiments.

2.3.1   *Design of single loop control*

The adjustment of controller parameters is solved by the following methods:

— Application of tuning rules based on the step response of the plant, as developed by Chien *et al.* [1].
— Calculation of parameters using the transfer function of the plant and approximation of the open loop transfer function by a IT_1 (integral plus time constant characteristic to obtain a dominating pole pair in two steps:
  (i)   compensation of large time constants by controller zeros;
  (ii)   determination of the amplification dependence of the overshoot of the command response [2].
— Calculation of controller parameters using integral criteria, e.g.

$$\int_0^\infty \left\{ e^2(t) + rk_s^2 \left[y(t) - y(\infty) \right]^2 \right\} dt = \text{Min!} \qquad (2.1)$$

where the control error is e, the weighting factor r, the static plant amplification k_s and the correcting signal y.

2.3.2 Multi-loop control

Here two methods of introducing additional controlled variables to reduce the influence of disturbances are presented:

— using a secondary controlled variable; and
— cascaded control.

In the case of the secondary controlled variable, the large time constants of the plant are compensated by an auxiliary controller, which is in parallel with the plant (see Figure 2.15). For cascaded control the faster work rate of the inner control loop (relative to the outer loop) reduces the influence of disturbances (Figure 2.16).

2.3.3 Two-variable control

A two-variable transfer element is a special case of a multi-variable transfer element. It has two input variables and two output variables. The general case of multivariable systems will be covered in Chapter 9

The input variables y_1, y_2 influence both output variables, and the output variables x_1, x_2 depend on both input variables (Figure 2.1). $G(s)$ is the transfer matrix between the input variables y_1, y_2 and the output variables x_1, x_2.

$$G(s) = \begin{bmatrix} S_{11}(s) & S_{12}(s) \\ S_{21}(s) & S_{22}(s) \end{bmatrix}$$

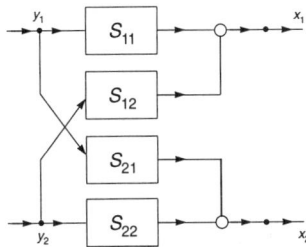

Figure 2.1   Two-variable transfer element and transfer matrix

Many technical multi-variable plants can be regarded as two-variable plants. The design of a controller for such plants includes the following tasks:

— designing in time or frequency domain (here the frequency domain is used);

- selecting a controller with complete, partial or no decoupling;
- finding suitable co-ordination between control and controlled variables;
- design of the main controllers and decoupling controllers;
- estimation of control behaviour, disturbance rejection, stability and integrity by simulation and in practice.

Figure 2.2 shows the structure of a control loop with plant S, decoupling controller K, main controller R, the vector of controlled variables *x*, disturbance variables *z*, command variables *w*, control variables *y* and control errors *e*.

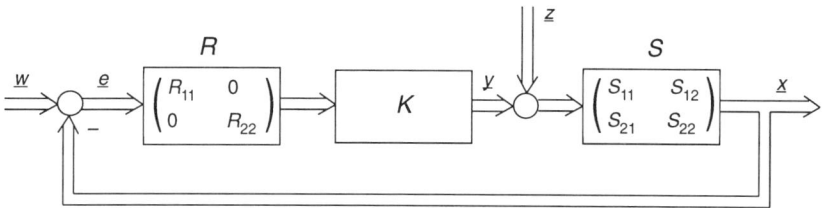

Figure 2.2   Two-variable control loop

Two-variable control without decoupling means

$$K = \begin{bmatrix} 1 & 0 \\ 0 & 1 \end{bmatrix} \quad \text{or} \quad K = \begin{bmatrix} 0 & 1 \\ 1 & 0 \end{bmatrix} \tag{2.2}$$

A completely decoupled control uses a decoupling controller

$$K = S^{-1}\begin{bmatrix} S_{11} & 0 \\ 0 & S_{22} \end{bmatrix} = \frac{1}{1-C}\begin{bmatrix} 1 & -S_{12}/S_{11} \\ -S_{21}/S_{22} & 1 \end{bmatrix} \tag{2.3}$$

with the coupling coefficient

$$C(s) = \frac{S_{12}(s)\,S_{21}(s)}{S_{11}(s)\,S_{22}(s)} \tag{2.4}$$

2.3.3.1   Design of the main controller without decoupling

Using the standard method for the resulting SISO plant (Figure 2.3), the main controller $R_i = R_{ii}$ has to work with a plant characterised by

$$S_{ii}^*\left(s\right) = S_{ii}\left(s\right) - S_{ij}\left(s\right)S_{ji}\left(s\right)\frac{R_j\left(s\right)}{1 + R_j\left(s\right)S_{jj}\left(s\right)} = S_{ii}\left(s\right)\left[1 - C\left(s\right)F_j\left(s\right)\right] \qquad (2.5)$$

$i,j = 1,2, j \neq i$. Equation 2.5 contains the command transfer function of the other loop with disconnected controller R_i,

$$F_j\left(s\right) = \frac{R_j\left(s\right)S_{jj}\left(s\right)}{1 + R_j\left(s\right)S_{jj}\left(s\right)} \qquad (2.6)$$

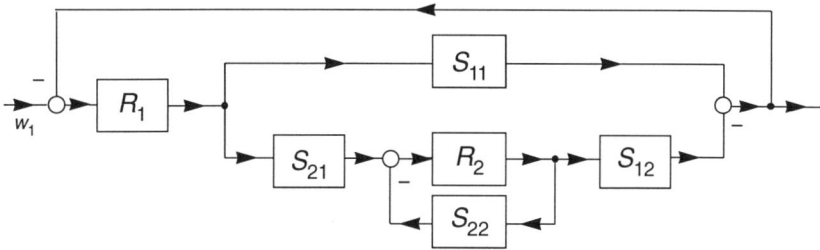

Figure 2.3   Controller R_1 with associated plant

For the design of the controller $R_i(s)$, the other controlled loop is supposed to be approximately unity, that is,

$$F_j\left(s\right) \approx 1 \qquad (2.7)$$

Therefore the controller is to deal with a plant transfer function

$$S_{ii}^*\left(s\right) = S_{ii}\left(s\right)\left[1 - C\left(s\right)\right] \qquad (2.8)$$

like a SISO system. In this case the coupling within the plant is considered by the coupling coefficient $C(s)$. It is useful to confirm the results by simulation. The characteristic polynomial for the stability check is

$$1 - C\left(s\right)F_1\left(s\right)F_2\left(s\right) = 0 \qquad (2.9)$$

with the command transfer function according to Equation 2.6.

2.3.3.2   Design of decoupled control
If a good plant model is available, the decoupling controller can be calculated from Equation 2.3. The result, given in Figure 2.4, is a controller with so-called P structure according to the transfer matrix in Figure 2.1. Because of the inversion of the transfer functions and the conditions for numerator and denominator degrees, the decoupling controller is often only approximately realisable.

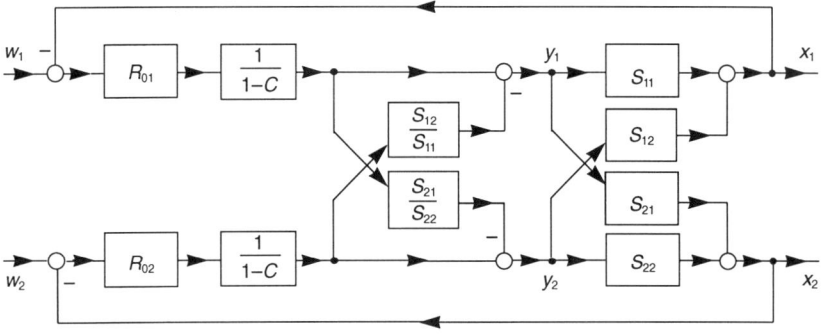

Figure 2.4   Control loop with decoupling controller in P structure

Another mechanism to build the decoupling controller is based on the V structure. Using the P-to-V conversion presented in Figure 2.5

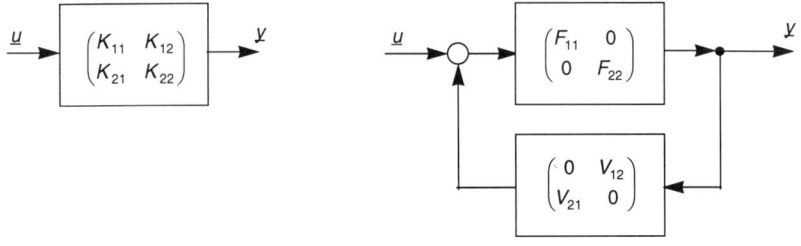

Figure 2.5   P-to-V conversion

and using the formulas

$$F = \det K \begin{bmatrix} 1/K_{22} & 0 \\ 0 & 1/K_{11} \end{bmatrix}, \quad V = \frac{1}{\det K} \begin{bmatrix} 0 & K_{12} \\ K_{21} & 0 \end{bmatrix} \qquad (2.10)$$

for the feedforward and feedback transfer, the simple decoupling controller in Figure 2.6 results.

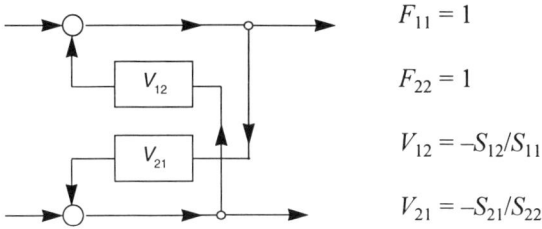

$$F_{11} = 1$$
$$F_{22} = 1$$
$$V_{12} = -S_{12}/S_{11}$$
$$V_{21} = -S_{21}/S_{22}$$

Figure 2.6   Decoupling controller in the V structure

Complete decoupling guarantees autonomous control, but the disturbance behaviour of such a system can be worse than that of a control system with partial or no decoupling.

2.3.4   Two- and three-level control

Many technological plants contain static non-linearities, e.g. saturation and dead zone. For economic reasons, industrial controllers also use non-linear elements. In many cases the control loop has the form of Figure 2.7 with separate static non-linear and linear dynamic transfer elements [3].

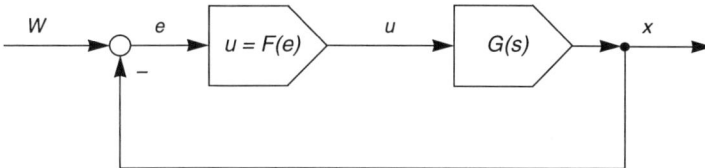

Figure 2.7   Non-linear control loop

An analogue simulation is useful to investigate the dynamics of the system, especially its stability behaviour.

2.3.4.1   Application of the method of describing functions
Under the precondition of low pass behaviour for the linear part and some simplifying assumptions (e.g. symmetry of the non-linear element, zero mean) this approach can be used to

—   prove the stability;
—   estimate the margin of stability;
—   calculate the amplitude and frequency of the oscillation.

The main idea is to describe the non-linear transfer by means of a function $N(A)$ of the input amplitude A as the relation between the fundamental oscillation of the output and the sinusoidal input.

Figure 2.8 shows a typical three-level element and the corresponding describing function $N(A)$.

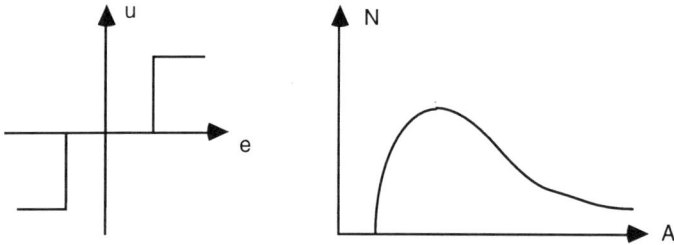

Figure 2.8   Non-linear element and corresponding describing function

The equation of harmonic balance

$$1 + G(jw)N(A) = 0 \tag{2.11}$$

enables a study of the existence of a limit cycle to be undertaken, together with its frequency and amplitude, which can be determined analytically, numerically or graphically.

2.3.4.2   Phase plane

For non-linear systems of second order, the phase plan is a very instructive way to represent dynamic behaviour. Representation in the phase plane can be achieved by:

— the method of isoclines; or
— the method of switching trajectories.

Isoclines help to sketch trajectories by beginning at a starting point and continuing in the given direction of isoclines, Figure 2.9. The field of isoclines is calculated from the second order state equations

$$dx_1 / dt = f_1(x_1, x_2), \quad dx_2 / dt = f_2(x_1, x_2) \tag{2.12}$$

$$\frac{dx_2}{dx_1} = \frac{f_2(x_1, x_2)}{f_1(x_1, x_2)} = c \tag{2.13}$$

Equation 2.13 describes the course of the isocline for a given inclination *c*.

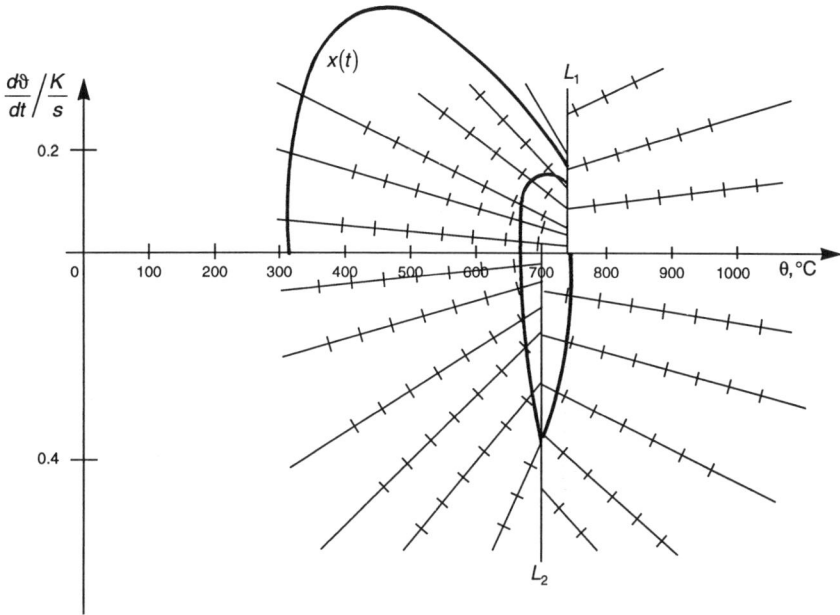

Figure 2.9   *Field of isoclines with marked inclinations, switching lines L_1, L_2 and trajectory x(t)*

If the non-linear element, such as a clipper or multi-level element, is step-wise linear, then the trajectories can be constructed by switching the solutions of the different linear fields. Switching lines separate the different fields (Figure 2.9).

Simulation is a useful method of checking the results.

2.4   Laboratory set-up (simulation tools)

The laboratory set-up consists of a special electronic analogue computer with a programmable signal generator, a control unit with different modules implementing typical transfer elements, an oscilloscope and an X-Y recorder (Figure 2.10). Control structures of different complexity can be built, as the different components are linked by wires according to a given block diagram. The programmable signal generator produces the necessary inputs for control and distortion. By means of the oscilloscope and X-Y recorder the signals can be observed at any point. Thus it is possible to investigate the effects of different control structures and transfer elements on the output signals. The use of the set-up promotes the understanding of control structures and helps to develop student's skills in measurement.

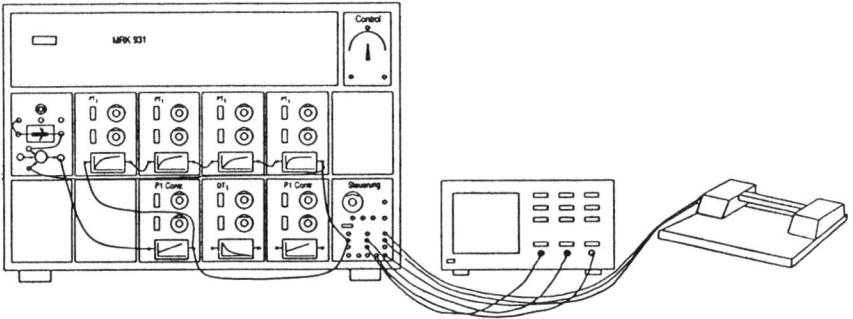

Figure 2.10   Laboratory set-up

For the simulation of plants, controllers and control loops, the following modules are principally necessary:

— lag elements of first and second order;
— oscillating elements;
— integrators and differentiators;
— amplifiers;
— two- and three-level transfer elements (non-linear controller);
— signal generators and control unit;
— X-Y recorder, oscilloscope.

as shown in Figures 2.10 and 2.11 and Table 2.1. The flexibility of the three-level transfer element is shown in Figure 2.12.

The set up makes possible a great variety of experiments.

Figure 2.12 shows the structure of an adjustable non-linear transfer element. The circuit of a PT2 transfer element (lag element with two time constants) is represented in Figure 2.13.

The programmable signal generator produces step, ramp, sinusoidal and random functions.

All outputs and inputs are short-circuit and voltage protected. Therefore arbitrary connection of inputs and outputs of the same module, and between different modules, will not damage the equipment. The operation range of the output voltage lies between −10 and +10 V.

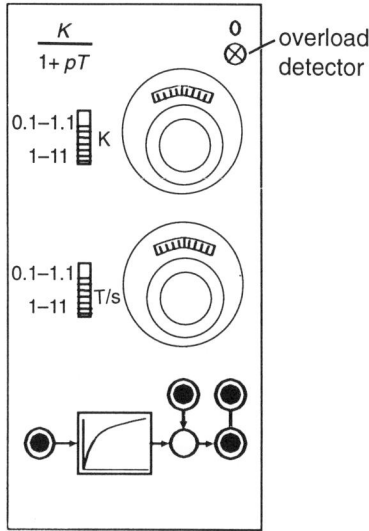

Figure 2.11   Front panel of lag element

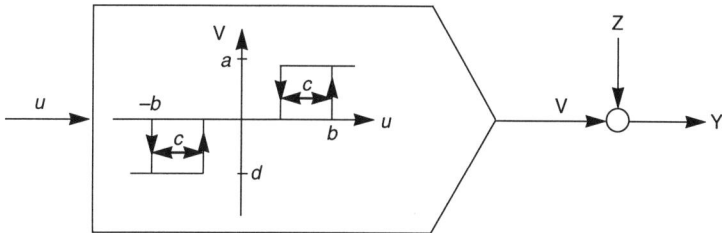

Figure 2.12   Non-linear transfer element
$a = 0 \ldots 10\,\text{V}, b = 0 \ldots 10\,\text{V}, c = 0 \ldots 10\,\text{V}, d = 0 \text{ or } -a$

The system operates in real time and repetition mode. The variables of interest are represented by the X-Y recorder and the oscilloscope, respectively. Time scaling by a factor of 1000 permits a 'permanent' representation by oscilloscope in the repetition mode.

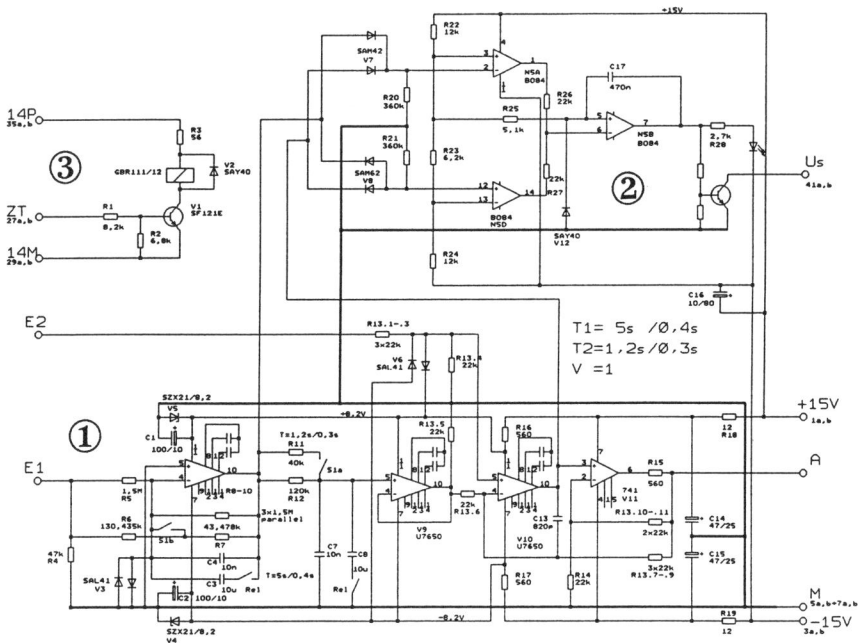

Figure 2.13   Circuit of the PT2 (two time constants) transfer element
1 main circuit for realising the transfer function
2 overload detector
3 change of operating mode (real-time and repetition)

2.5   Suggested experiments and problems

The experiments include:

— connecting the transfer elements according to the desired block diagram;
— measuring transient functions of controlled variables and manipulating variables in different control structures;
— investigation of the influence of system parameters on transient and stability behaviour;
— tuning controller parameters;
— discussion of the results.

Table 2.1   *Linear transfer elements available*

	$y = v + z$ $V(s) = G(s)\,U(s)$	
$G(s)$	T (s)	k or D
k		0 ... 1 0 ... 10
$1/(sT)$	0.1 ...1.1 1 ... 11 10 ... 110	
$k/(1+sT)$	0.1 ...1.1 1 ...11	0.1 ... 1.1 1 ... 11
$(1+sT_V)/(1+sT)$	T_V: 0.1 ... 1.1 1 ... 11 T: 0.1 ... 1.1	
$ksT/(1+sT)$	0.1 ... 1.1 1 ... 11	0.1 ... 1.1 1 ... 11
$0.5/(s^2T^2+2DsT+1)$	0.1 ... 1.1 1 ... 11	0 ... 1
$k[1+1/(sT_N)]$	0.1 ...1.1 1 ... 11	0 ... 10
$k[a+1/(sT_N)+sT_V/(1+sT)]$ $a = 0$ or 1	T_N: 0.1 ... 1.1 1 ... 11 10 ... 110 T_V: 0 0.1 ... 1.1 1 ... 11 $T = T_V/20$	0 ... 10

The following have been selected from the great variety of experiments possible.

For the given control loop in Figures 2.14, 2.15 and 2.16, the response of the correcting variable and controlled variable should be measured to a step of reference and/or disturbance variables, showing their dependence on different parameters of the PI controller.

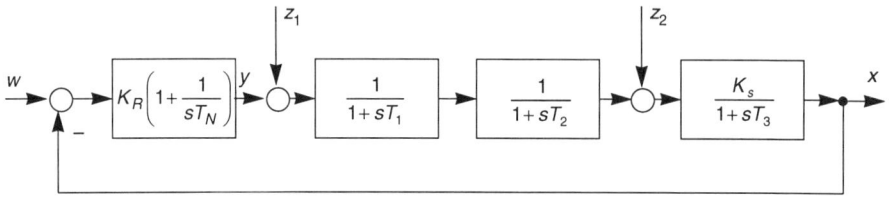

Figure 2.14   Single loop control
    $K_S = 2$, $T_1 = 5$ s, $T_2 = 1.2$ s, $T_3 = 0.2$ s

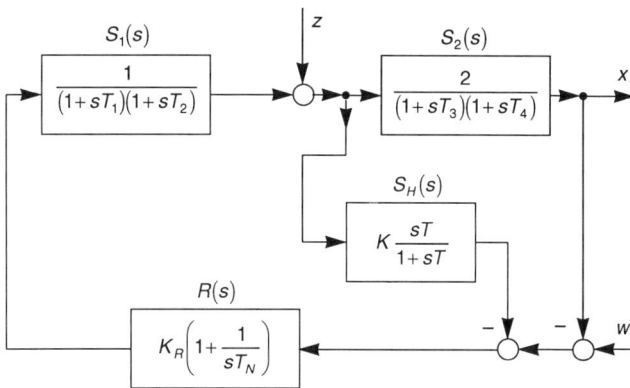

Figure 2.15   Control with auxiliary corrector
    $T_1 = 0.4$ s, $T_2 = 0.3$ s, $T_3 = 2$ s, $T_4 = 1$ s

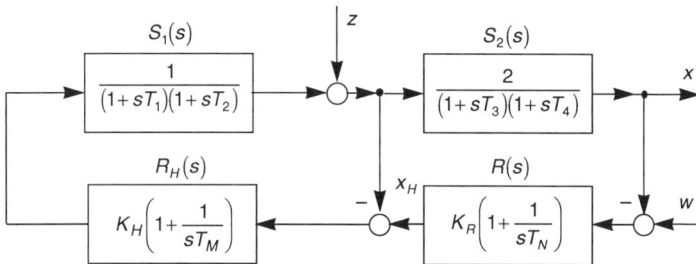

Figure 2.16   Cascaded control

A simplified model of a real technical plant, the bottom of a distillation column, is given in Figure 2.17. The input variables are vapour stream y_H and output material flow y_A, and the output variables are the temperature x_T and the level x_S.

The design of the controller mainly depends on the definition of the input and output vectors y and x, respectively.

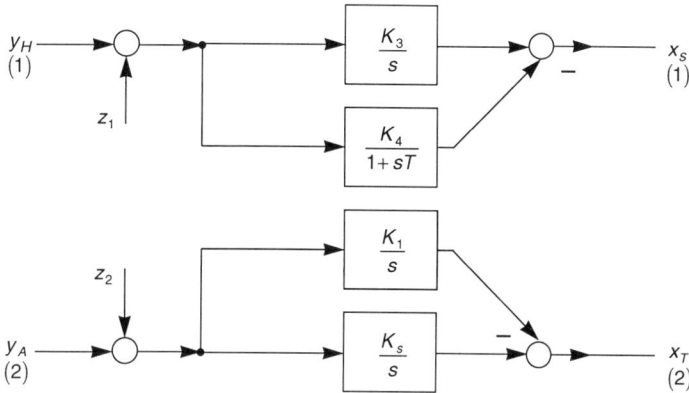

Figure 2.17   Plant model of the bottom of distillation column

The following problems are set:

— co-ordination of the regulating and controlled variables, respectively;
— design of a controller without decoupling and with complete decoupling; and
— measurement of the command and disturbance behaviour.

Figure 2.18 shows an example of temperature control of a plant of second order with a two-level controller.

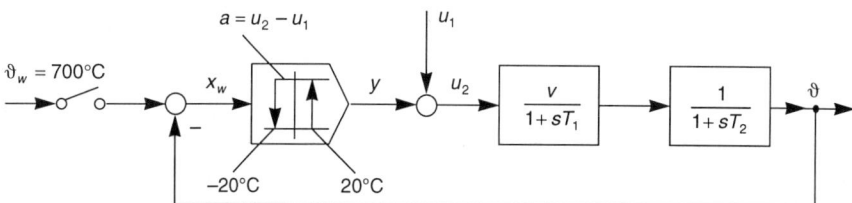

Figure 2.18   Temperature control
u_1 basic heating power, a additional heating power

Here the problem is to compare theoretical with measured results, especially in the behaviour of the transient trajectory.

2.6   Illustrative results

An essential goal of these experiments is to become acquainted with the step responses of different control systems in the cases of reference control and disturbance rejection [$w(t)$ and $z(t)$ respectively].

2.6.1   Single-loop and multi-loop control

Results from the suggested experiment are presented in Figures 2.19, 2.20 and 2.21.

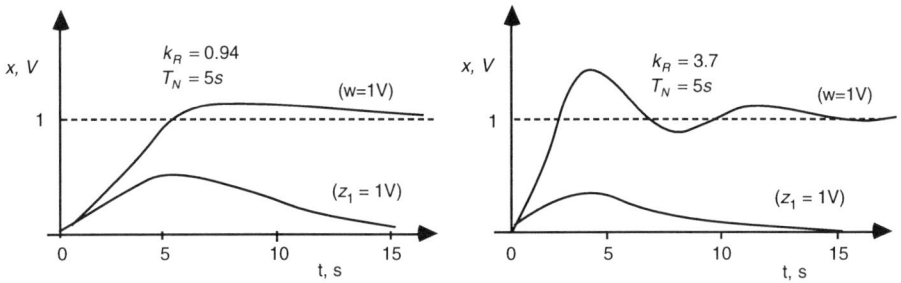

Figure 2.19   Simple single-loop control according to Figure 2.14

2.6.2   Two-variable control

Based on the plant model given in Figure 2.17, the calculated decoupling controller and main controller give the resulting control system in Figure 2.22.

Figure 2.23 shows the experiments with reference step changes $w_1 = 1$ V and $w_2 = 0$.

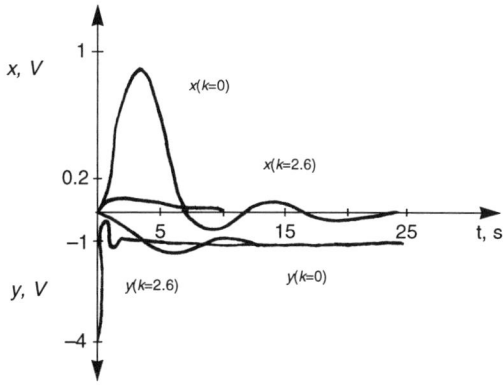

Figure 2.20 *Control with auxiliary corrector (Figure 2.15)*
<u>K = 0</u>: z = 1 V, K_R = 0.65, T_N = 2 s
<u>K = 2.6</u>: z = 1 V, K_R = 1.3, T_N = 0.4 s, T = 2.3 s

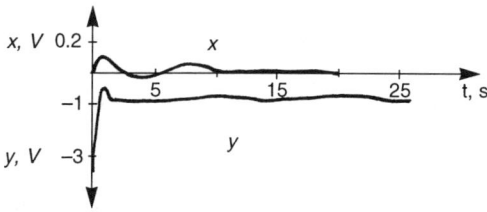

Figure 2.21 *Cascaded control (Figure 2.16)*
z = 1 V, K_R = 1.05, T_N = 2 s, K_H = 2.6, T_M = 0.4 s

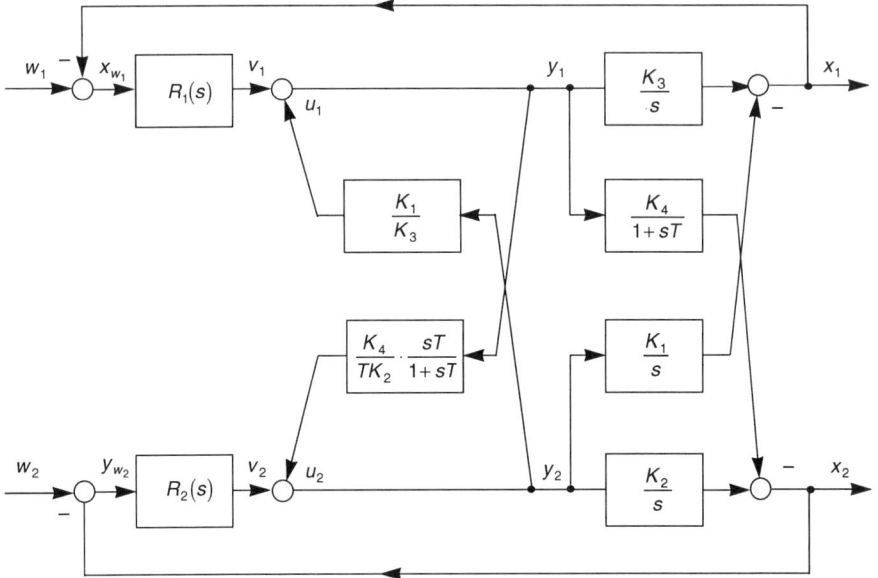

Figure 2.22   Two-variable control
$K_1 = 1.28\,\text{s}^{-1}$, $K_2 = 1.20\,\text{s}^{-1}$, $K_3 = 0.40\,\text{s}^{-1}$, $K_4 = 0.32$, $T = 1\,\text{s}$

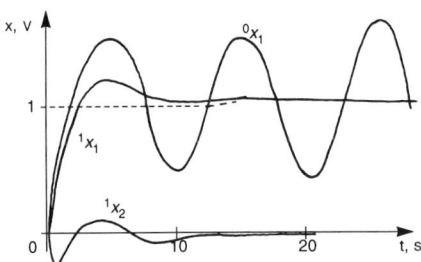

0_{x1} without decoupling
$V_{12} = V_{21} = 0$
$K_{R1} = 1{,}7$; $K_{R2} = 0.57$
$T_{N1} = 1.3\,\text{s}$; $T_{N2} = 1.3\,\text{s}$

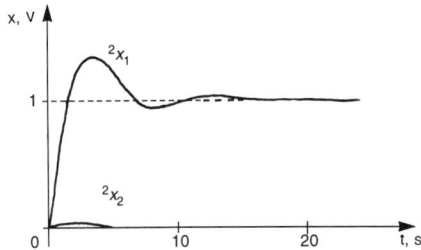

1_{x1}, 1_{x2} design according to Equation 2.8
$V_{12} = V_{21} = 0$
$K_{R1} = 3$; $K_{R2} = 1$
$T_{N1} = 10\,\text{s}$; $T_{N2} = 10\,\text{s}$

2_{x1}, 2_{x2} design with de-coupling,
$V_{12} = V_{21} = 0$
$K_{R1} = 1{,}7$; $K_{R2} = 0.57$
$T_{N1} = 1.3\,\text{s}$; $T_{N2} = 1.3\,\text{s}$

Figure 2.23   Command behaviour of the two-variable control system

2.6.3 Non-linear control

Figure 2.24 shows the transient of the controlled variable ϑ (300–700 °C), obtaining a limit cycle in the time domain and the phase plane with the following data:

$v = 1,$ $\qquad$ $\vartheta_w = 700\,°C = 7\,V,$ $\qquad$ $c = 40\,°C = 0.4\,V,$
$T_1 = 1\,s,$ $\qquad$ $a = 5\,kW = 5\,V,$
$T_2 = 1\,s,$ $\qquad$ $u_1 = 3\,kW = 3\,V.$

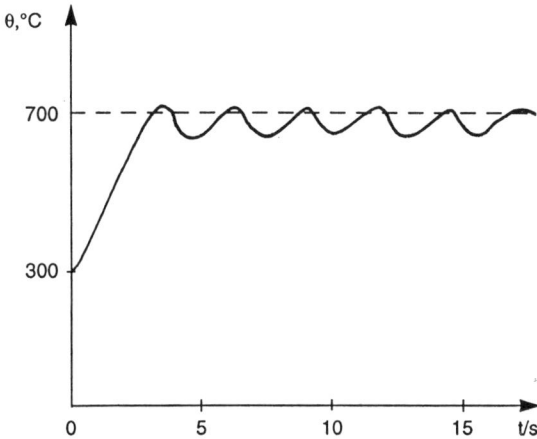

Figure 2.24   *Transient of temperature control in the time domain (phase plane according to Figure 2.9)*

2.7 Conclusions

Such simulations show that:

(a) the design of the controller depends on the type and location of input signals for command and disturbance, whereas a control loop sufficiently suppresses only those disturbances for which it has been designed;

(b) high control quality demands high power and rapid rate of change of the correcting variable;

(c) normally many solutions exist for any one problem and the results are often equivalent.

Simulation is an important step in the design process. Therefore an early and adequate introduction to simulation is important to the student. Use of an analogue computer can give a more accurate approximation of the physical reality, together with added clarity when measuring the interesting variables, than digital computer simulation provides.

2.8   References

1   CHIEN, K. L. HRONES, J. H. and RESWICK, J. B.: 'On the automatic control of generalized passive systems', *Trans. ASME*, 1952, **74**, pp.175-185

2   REINISCH, K.: 'Application of a model control loop to obtain design rules for linear control systems and to calculate of system parameters', *MSR*, 1962, **6**, pp. 245-251 (in German)

3   COOK, P. A.: 'Nonlinear dynamical systems' (Prentice/Hall, Englewood Cliffs, NJ, 1986)

Chapter 3

Classic controller design

P. Bikfalvi and I. Szabó

3.1 Introduction

Classic controller design is most commonly used in automatic engineering. Standard industrial controllers use different variants of PID-control or lead-lag compensation. Digital controllers and PCs with industrial interface cards have widened the field towards more advanced control algorithm implementation, even if some of them are based on classic controller design.

A review of theoretical and practical approaches to classic controller design methods is presented here, together with possibilities for digital realisation. Some special PC software, developed to implement some common digital algorithms, is also presented. The software is mainly for educational purposes, but may also be helpful in automatic system research and development.

3.2 Control problem (motivation)

Many control problems arise from the design of engineering systems. Such problems are typically large-scale and fuzzy. Common examples include the design of power plants, chemical processes, metallurgical processes, industrial robots, aircraft and biomedical systems.

On the other hand, control theory deals with small-scale, well-defined problems. A typical problem is that of the design of a feedback law for a given single-input single-output system, which is described by a differential equation with constant coefficients, so that the closed-loop system has given poles. The system theory related to this problem usually uses methods based on the transfer function (Laplace transform), the frequency domain or, sometimes, the time domain.

Among the many difficult problems relating to control system design, it is important to note the relationship between process design and control design. In the early days of automation, control systems were often introduced into existing, or already designed, processes to improve their operation. This still happens today, even if the process and the regulator are designed together as they are in many cases. The resulting drawback is that specific regulator structures, actuators, sensors and estimators need to be designed to suit the specific situation. The regulator's design is therefore strongly influenced by the effort that has been put into the process design. In many cases it is not economical to make much effort, especially if a standard controller will suit the system.

Classic automatic control system design can be understood simply as a problem of selecting the type of regulator and calculating its parameters to ensure the desired static and dynamic behaviour of the controlled system. The specific features of such control systems are:

- division of the system into two components: the plant, including the actuator and the sensor, and the controller (the inertia of which is negligible in comparison with that of the plant);
- use of standard controllers.

Thus the generalised design problem reduces to the selection of a suitable controller, which would ensure the desired control law, and adjustment of the controller's parameters to suit the plant's required dynamic and static responses.

Such control loops mostly use a standard, all-purpose regulator with adjustable parameters. The appropriate parameters are found by using a variety of defined tuning rules. Modern regulators use microprocessor-based computer control. This increases the flexibility of realisable control laws, operation is easier and the controller is cheaper and more reliable.

Control system design is a compromise between many factors. The following issues must always be considered: model uncertainty, command signals, actuator saturation, disturbances, state constraints, regulator complexity and cost. Few design methods consider all of these factors. It might be argued that a good design is almost never the best one. Therefore, investigation through analysis and/or simulation is also necessary and must be carried out with care.

Classic design first sets the control system configuration. Figure 3.1 shows structural diagrams of the most popular controlled systems.

In the first case (Figure 3.1*a*) the most commonly used cascade (series) compensation is shown. The compensator (regulator) is incorporated into the control loop. The open-loop transfer function of the compensated system is:

$$G(s) = G_c(s) \cdot G_p(s)$$

where: $G(s)$ is the resulting open-loop transfer function;
       $G_c(s)$ is the transfer function of the cascade (series) regulator;
       $G_p(s)$ is the transfer function of the uncompensated process.

In parallel compensation (Figure 3.1*b*) the regulator is situated in a special feedback loop. In this case:

$$G(s) = \frac{G_1(s) \cdot G_2(s)}{1 + G_2(s) \cdot G_c(s)}$$

where:

$G(s)$ is the resulting open-loop transfer function;

$G_1(s)$ is the transfer function of the outside compensating feedback loop part of the system;

$G_2(s)$ is the transfer function of the inside compensating feedback loop part;

$G_c(s)$ is the transfer function of the parallel regulator.

Formally, parallel compensation can always be selected in such a way as to obtain the same result as in series (cascade) compensation, and vice versa. In practice, the choice between these compensating structures depends on the operational characteristics of the component functional elements.

Both of these methods use feedback. Their advantage is that sensitivity to disturbances and to system parameter variations is reduced. Feedback is most effective when the process dynamics need a high bandwidth. For this reason most valuable design methods have been developed using these systems.

The third diagram (Figure 3.1*c*) shows a special structure and is mostly used for measurable disturbance compensation. The regulator is part of a supplementary feedforward loop, along which the signal to be compensated is introduced into the system. This compensation method differs substantially from the two above in its working principle. The structural diagram shows that this compensation cannot affect the stability of the closed-loop system, but improves the system's performance by reducing the dynamic error related to the disturbing action that is to be compensated. It is easy to demonstrate [3] that by selecting the regulator transfer function with the condition:

$$G_c(s) = \frac{1}{G_2(s)}$$

the dynamic error of the system can be eliminated, irrespective of the nature of the action $X(s)$. Thus, the design of a feedforward compensator is in essence a calculation of the inverse of a dynamic system.

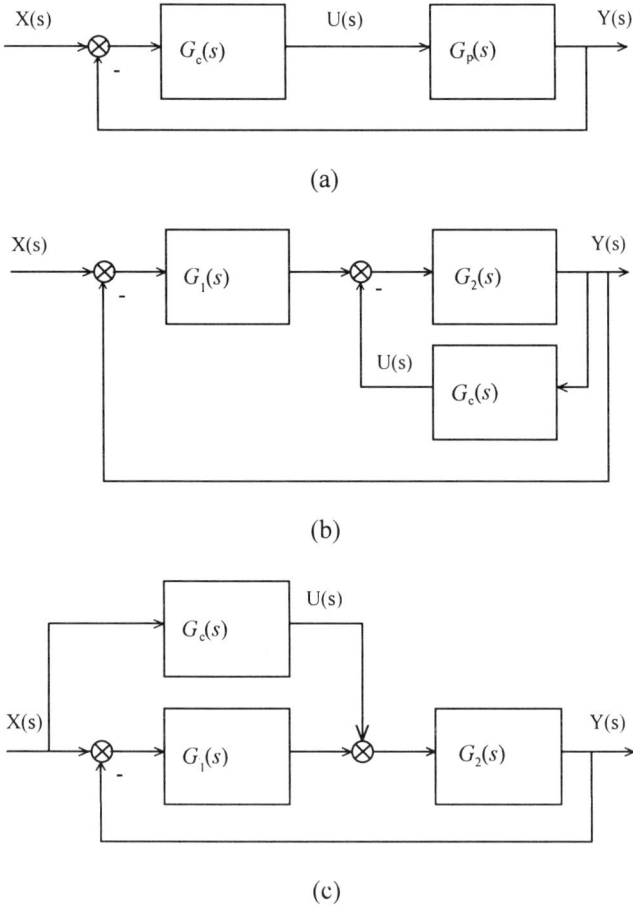

(a)

(b)

(c)

Figure 3.1   Popular structural diagrams of controlled systems
   a   cascade (series) compensation
   b   parallel compensation
   c   feedforward compensation

The use of this method involves many technical difficulties in analogue realisation, which often prevent complete elimination of the dynamic error. This is because the transfer function $G_2(s)$ represents (as a rule) an integrating, inertial or oscillatory element, or a combination of them (a real physical element). As a consequence, the compensating element transfer function should correspond to differentiating elements, whose practical implementation is difficult to achieve.

The advantage of feedforward, compared with feedback, is that corrective action may be taken before the disturbance has influenced the variables. Since feedforward is an open-loop compensation method, it requires a good process model. The use of feedforward control is better applied to digital control, where this is widely permitted.

The compensation methods described above are not mutually exclusive. On the contrary, in many cases, to achieve the desired performance a combination of feedforward and feedback compensation must be applied. A common example is shown in Figure 3.2, where $G_{c1}(s)$ and $G_{c2}(s)$ are the feedforward and feedback compensators, respectively.

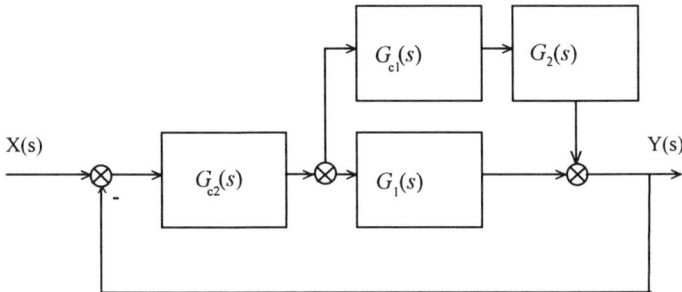

Figure 3.2   Combined feedback and feedforward compensation

Selecting a suitable control system configuration strongly depends on the designer's skill and experience. Specifications can be overly contradictory or too strict which can lead to unsatisfactory or unrealisable solutions.

Classic controller design methods will give always an approximate solution, which nevertheless can be considered the best. For example, assuring the desired phase margin or gain margin for a controlled system does not mean that this solution is optimal if a different performance indicator is used.

In everyday practice, automatic control systems use very different regulators or compensators. Analogue realisations can be simple RC-networks (low-pass, high-pass or band-pass filters) or may include active elements (operational amplifiers) too. In most cases, standard industrial controllers implement different versions of PID control. If they are microprocessor-based, translation into a digital form of the PID control law is implemented within them. A very short sampling interval is also characteristic.

3.3   Technical approaches to classic regulator design

There are many dedicated methods used to estimate regulation performance and to design regulators (compensators). Regulation problems are often solved by feedback. Feedforward techniques are also useful to reject the measurable disturbances or in servo-system design.

This section deals only with single-input single-output, linear system control design methods, especially those that belong to classic control design. The brief overview of each method is given for introductory purposes — the reader is advised to study the relevant literature in more detail.

3.3.1   Graphical analytical methods

Controller design based on frequency-response

In its early period, the theoretical development of control system design was characterised by Nyquist, Hall, Nichols and Bodè. They developed graphical analytical methods of design that are consecrated in the literature with their names. Easy to understand and apply, all these methods are commonly used. Many computer-aided tools were subsequently developed from them.

The methods are based on the frequency response of an open-loop system and do not require the solution of the model's differential equations. Estimation of the transient and the steady-state performances of the closed-loop system, such as magnitude and phase margin, bandwidth, cut-off frequency, oscillation index, etc., is also possible. In consequence, compensators designed this way mostly have leading, lagging or PID effects, or a combination of these.

Detailed description of these methods can be found in almost any process control theory book published in the 1960s or even 1970s [2,3].

Controller design based on root locus

Root locus is a graphical analytical method for analysis and synthesis in the *s*-plane for linear control systems. Its main advantage results in obtaining the time and frequency domain specifications directly from the characteristic polynomial roots (poles) location in the *s*-plane. A disadvantage of the method is that the design procedure is a step-by-step approach. A solution is easily achieved if computer aid is available.

We can mention particularly the work of Evans and Truxal, but detailed descriptions of the method can be found in most of the books mentioned above.

3.3.2   Analytical methods

Controller design based on the transfer function

A very suitable way to study linear control systems is based on the concept of the transfer function. It is then natural to apply methods that allow us to change this transfer function.

One of the most commonly used design methods is that of pole placement. The main idea is to determine the parameters of a general form regulator (coefficients of polynomials), so that the closed-loop system has the desired properties. A detailed description of the method appears in many control theory books from the last two decades, many of the latter deal especially with digital control design [4–6].

The method can be presented in a very simple form as follows. Let us consider a feedback control (Figure 3.1*a*) and $G_0(s)$ the transfer function of the desired closed-loop system. In this case the desired open-loop transfer function will be:

$$G(s) = \frac{G_0(s)}{1 - G_0(s)}$$

The transfer function of the controller is given by:

$$G_c(s) = \frac{G_0(s)}{1 - G_0(s)} \cdot \frac{1}{G_p(s)}$$

The method can be applied to both continuous and sampled linear systems. The design method usually uses an approximate process model and many of the subjective judgements of the control engineer can be incorporated into the desired behaviour model.

Pole placement is a general approach in single-input single-output, process control design. Many other design methods, such as root locus, dipole compensation, Smith predictor, Dahlin control algorithm, dead-beat control, model algorithmic control etc., can also be considered as particular aspects of the pole placement design. Two of these will be presented as suggested experiments.

Controller design based on time domain response

Usually the performance of a controlled system is expressed in the time domain by the specification of its step or impulse response. These types of specification include rise time, overshoot, settling time, steady-state error etc. In most cases the design problem is solved using analytical methods.

The performance criterion that satisfies the desired time specifications is usually expressed as a time integral optimisation criterion. The synthesis problem is formulated to minimise this time integral, where the integrand is commonly a quadratic loss function. The mathematical problem is well known in technical literature as the minimum variance control problem for the input-output approach, or the linear quadratic problem for the state-space approach. A detailed description of this issue can be found in References 4–7.

This is a very modern design technique but serious difficulties may appear in the analytical solution process. Also, there is no general rule for choosing the form of the integrant function. In many cases only computer aid can help.

Smith predictor design

The Smith predictor is a special design method for dead-time process control. The block diagram in Figure 3.3 shows the working principles of such a regulator.

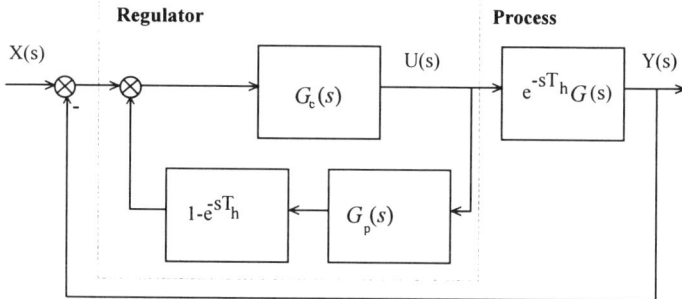

Figure 3.3   Smith predictor: working principle

The method can be useful in the compensation of large dead-time processes, where the part of the process with the dead-time (time delay) cannot be separated but the dead-time itself may be precisely determined. Using this method, the regulator does not compensate the dead-time delay, but improves the performance of the part of the process without dead-time.

It is easy to note from Figure 3.3 that the regulator contains the time delay, so it is necessary to store the signals. This is difficult using analogue implementation, but easy using a digital computer. The time delay can be represented by a vector, which is shifted at each sampling instant.

3.3.3   Controller design based on rule-of-thumb methods

Difficulties in precise estimation of process parameters, the non-linear behaviour of real processes, the stochastic character of certain disturbances and the difficulties in finding analytical solutions, all lead to limited usefulness of analytical or graphical-analytical methods. Another important reason for introducing empirical methods of controller design is the fact that standard controller tuning and performance checks must frequently be done locally in the working process. Even if analytical controller design is available, final 'fine-tuning' is necessary almost every time.

Empirical methods have been developed for tuning standard controllers. For a PID controller this means primary parameter setting, namely the proportional factor (or proportional band), the integral time constant and the derivative time constant.

The relevant literature gives many examples of such rule-of-thumb methods. This demonstrates again the fact that controller design is a very specific domain, where both theoretical knowledge and practical experience are important. Among these design methods, two have been accepted as more general: the step response method and the ultimate sensitivity (Ziegler–Nichols) method. A detailed description is available in almost any control theory or automatic control book [2,4,5].

3.4   Laboratory set-up

The basic laboratory equipment available in almost any process control laboratory (signal generator, DC current or voltage source, technological models, analogue or digital computers) is very useful. However, new software will need only an IBM PC (or compatible) and, optionally, a common data interface card (PCL 718).

Practical experiments can be done using real working models or process simulators (analogue or digital computers) with corresponding interfaces or, lately, the software may have its own demonstration options (if the interface card or the real model is missing).

3.5   Controller simulation software

The alternative laboratory set-up is, in fact, a controller simulation program. This is PC-implemented, real-time operating software for a standard industrial controller. The software was developed by the authors and is used in their teaching.

The recommended minimal hardware configuration is:

— IBM PC 80286 AT (16 MHz clock speed) or above;
— VGA display;
— Microsoft Mouse (driver installed);
— MS DOS 3.0 (or above);
— PCL 718 commercial data acquisition card (optional).

The software simulates the physical functions of a standard industrial controller, but it offers opportunities for higher control algorithm implementation too. Once started, the screen display shown in Figure 3.4 appears.

The basic state of the controller at start-up or reset is:

— manual mode of operation;
— set-point variable (SP) and manipulated variable (OUT) are set to zero;
— no control algorithm installed (TYPE = NONE);
— no filtering for the process variable (FILTER = OFF);
— no active alarms (ALARMS LED does not blink);
— 20 ms sampling period;

— direct mode of operation;
— graphic display cleared.

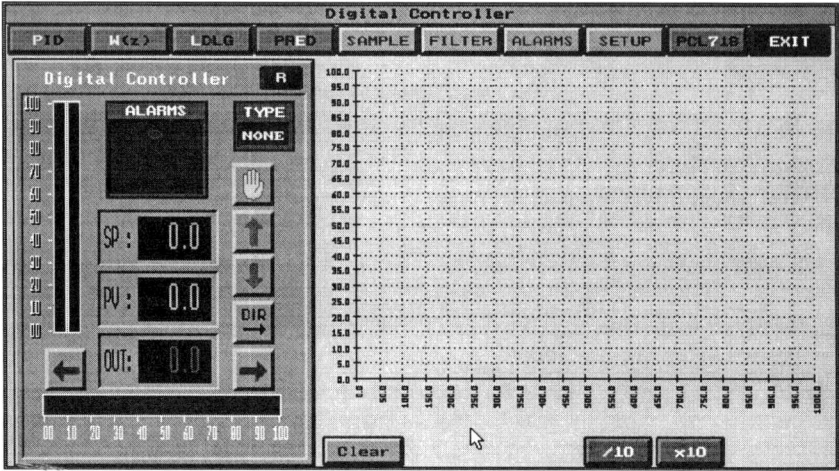

Figure 3.4   The digital controller display at start-up

If the PCL-718 card is installed and the signal corresponding to channel 1 of the A/D converter is connected, this appears in blue as the process variable (PV) value in digital form (PV display) and as a bar graph (left side vertical bar).

The manual operating mode permits the controller manipulated variable (OUT) setting, by clicking with the left mouse button on the ← or → push-buttons or pressing the ←,→ (left, right) buttons on the keyboard. The corresponding value is displayed in digital form and is shown by a cursor on the horizontal analogue display (red).

Clicking with the left mouse button on ✋ (MANUAL operation mode), the Ⓐ button appears instead and switching to AUTOMATIC mode occurs. (Clicking on it again, returns to MANUAL operating mode. The same effects can be obtained with the space-bar on the keyboard.) In automatic mode of operation the control algorithm set previously is executed. When either manual or automatic operation mode is installed, set-point variable (SP) setting is possible by clicking on the ↑ or ↓ buttons or pressing the ↑,↓ (up/down) buttons on the keyboard. The corresponding value is shown in digital or analogue (right side vertical bar graph) form in green.

Towards the top of the computer display, there are ten functional buttons, for the following different settings. Selection is possible by clicking on them with the left mouse button, or choosing the respective highlighted letter from the keyboard. Hereafter, only operation with the mouse is explained.

3.5.1   Sampling time setting

By clicking with the mouse on the SAMPLE button, a dialogue box opens for sampling time setting. A sampling period from 20 ms to 20 s, with a division of 1 ms, can be chosen. Detailed considerations on how to select sampling interval for different controller realisations can be found in References 4–6.

3.5.2   Filtering

A first order lag filtering of the process variable is possible by clicking on the FILTER button. A dialogue box opens and the filter time constant can be set. The filter must be switched ON. The default values at start-up are: time constant of 1 s (1 Hz cutting frequency), filtering OFF.

3.5.3   Demonstration process model operation

By clicking on the PCL718 button, this switches to DEMO showing that the controller is in demonstration operation mode. In this case, process variables communicate with one of installed demonstration process models. Clicking again on this button, the PCL718 button returns, which means that the controller is (or could be) linked to a real process via the PCL-718 interface card.

3.5.4   Set-up menu

Clicking on the SETUP button, makes a dialogue box appear depending on the operation mode selected above. If the demonstration operation is chosen, selection from three process models is possible:

- PT1TH — first-order process with dead-time, of the form:

$$G(s) = \frac{K_p e^{-sT_h}}{T_1 s + 1}$$

- PT2 — second-order, over-damped process, of the form:

$$G_p(s) = \frac{K_p}{T_1 T_2 s^2 + (T_1 + T_2)s + 1}$$

- PT2D — second-order, under-damped process, of the form:

$$G_p(s) = \frac{K_p}{T^2 s^2 + 2\zeta Ts + 1}$$

Parameters for each process model can easily be changed. If the PCL-718 card is installed, specific parameters of the interface card can also be set.

3.5.5   Alarms setting

By clicking with the mouse on the ⎢ALARMS⎢ button, a dialogue box opens where the low and high level of alarm signals can be set for the set-point, process variable, deviation and manipulating variable. If any of the configured alarms occur, these will be shown in the corresponding window of the front panel, and the alarm LED will blink.

3.5.6   Graphic menu

On the right hand side of the controller a graphic display appears. It is useful for graphical following of the set-point (green) variable, the process variable (blue) and the manipulating variable (red) values. Clicking on the ⎢/10⎢ and ⎢x10⎢ push-buttons allows the time interval to be displayed to be changed from 100 to 10,000 times the sampling period (from 2 s to approximately 50 h). Clicking on ⎢Clear⎢, deletes the trends and the graphic display begins from the left hand side again. Graphics appear continuously only in automatic working mode.

3.5.7   Control algorithm set-up

There are four different control algorithm types implemented in order to enable experiments with most of the well-known controller compensation laws. Controller tuning means filling in some set-up tables, including the corresponding selection of the sampling period.

PID control
In everyday practice one can find many different realisations of PID control. This program allows choice from three of them, namely:

- PID-A, of the form:

$$out = K_p \left(1 + \frac{1}{T_i s} + T_d s\right) \cdot dev$$

- PID-B, of the form:

$$out = K_p \cdot \left(1 + \frac{1}{T_i s}\right) \cdot \left(\frac{1 + T_d s}{1 + 0.1 T_d s}\right) \cdot dev$$

- PID-C, of the form:

$$out = K_p \cdot \frac{1}{T_i s} \cdot dev - K_p \cdot \left(\frac{1 + T_d s}{1 + 0.1 T_d s}\right) \cdot pv$$

where: *dev*   is the Laplace transform of the deviation variable;
          out   is the Laplace transform of the output variable;
          pv   is the Laplace transform of the process variable.

The deviation variable (DEV) depends on DIRECT 🔳 or REVERSE 🔳 operation mode, which means:

$$dev = pv - sp, \text{ or}$$
$$dev = sp - pv$$

respectively, where *sp* is the set-point variable.

Direct or reverse operation is easily selected by clicking with the mouse on the relevant switch push-button located on the controller front panel, or by pressing on the Tab (tabulator) key.

For each control rule, the corresponding dialogue box enables setting of the tuning parameters K_p, T_i, T_d (for example see Figure 3.5). This can easily be done for any of the selected control laws.

Figure 3.5   Dialogue box for PID controller parameter setting

Lead-lag control
This is one of the most common compensation types used in feedback and feedforward control. Its transfer function is:

$$G_x(s) = \frac{Y(s)}{U(s)} = K_p \cdot \frac{T_d s + 1}{T_i s + 1}$$

where: $Y(s)$ is the Laplace transform of the output variable (*out* in our case);
$U(s)$ is the Laplace transform of the input variable (*dev* in our case).

Clicking on the �**LDLG** button opens a corresponding dialogue box so that appropriate parameters (K_p, T_i, T_d) can be set.

Predictive control
A first-order Smith predictor algorithm is implemented. It is to be applied to first-order dead-time processes of form:

$$G_p(s) = \frac{K_f \cdot e^{-sT_h}}{T_f s + 1}$$

and leads to desired behaviour of form:

$$G_0(s) = \frac{1}{\dfrac{T_f}{K_m} s + 1} \cdot e^{-sT_h}$$

where K_m is the only tuning parameter and has practical values of between two and ten.

Clicking on the **PRED** button opens the corresponding dialogue box and suitable parameters can be set. Of course, parameter setting needs process parameters too.

General digital algorithm
The general impulse transfer function for a digital controller of the form:

$$G_c(z) = \frac{\displaystyle\sum_{i=0}^{n} a_i \cdot z^{-i}}{1 + \displaystyle\sum_{i=0}^{n} b_i \cdot z^{-i}}$$

is implemented, where the a_i and b_i coefficients must be established during the design procedure. The program permits implementation of an error cascade controller of regressive order up to $n=4$.

Clicking on the **H(z)** button, gives a specific dialogue box which permits coefficients to be set. Default values are zeroes.

3.5.8   Exiting the program

Exiting the program is only possible from MANUAL mode of operation, by clicking with the mouse on the **EXIT** button.

3.6   Suggested experiments

3.6.1   PID control experiment

During the experiment the following procedures have to be followed:

—parameter tuning for demonstration process models;
—qualitative analysis of performance for different tuning rules;
—analysis of control robustness for process parameter changes;
—analysis of influence of tuning parameter modification;
—parameter tuning for real continuous processes (if laboratory models or PCL-718 card available).

Analysis of disturbance rejection is also possible.

3.6.2   Lead-lag control experiment

During the experiment the following procedures have to be followed:

—parameter tuning for demonstration process models;
—qualitative analysis of the influence of the tuning parameter change on performance.

3.6.3   Predictive control experiment

During the experiment the following procedures have to be followed:

—design of a first-order Smith predictor;
—influence of tuning parameter (K_m) change: experiments on demonstration examples and/or real processes.

3.6.4 Digital cascade controller experiment

During the experiment the following procedures have to be followed:

—Dahlin–Higham algorithm design;
—dead-beat algorithm design;
—model algorithmic control design.

Demonstration examples or real laboratory models (if available) will confirm the experimental environment. Evaluation of the results is always necessary.

3.7   Illustrative results

Figure 3.6 illustrates some results obtained on the demonstration process examples given. Working with real processes (or models) in the laboratory environment, the teacher will have the task of linking the tool to the particular laboratory equipment available.

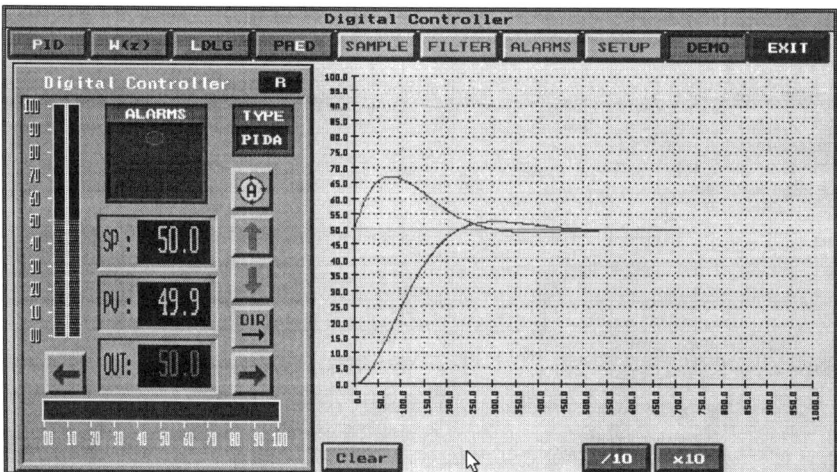

Figure 3.6   Illustrative working example of the controller

3.8 Conclusions and extensions

Classic controller design incorporates much of the basic knowledge that every control engineer must possess. The ability to teach it well, to make it easier to understand for students, is therefore important. The PC-implemented controller presented here is a useful tool and positive reactions from students demonstrate its advantages.

Further development of the software is directed towards implementation of a fuzzy PI-controller. This is in progress at the Department of Process Control, University of Miskolc, Hungary. Extensions are also needed to realise general pole placement design in input-output approach, state observer design and state control design in the state-space approach, but implementation of these methods on a PC needs different concepts and probably another software environment.

3.9 References

There are hundreds of books and articles which deal with the classic methods of controller design. These books and articles mostly relate to the problems of control theory or automatic process control. The reference list here is merely an extremely small selection of representative works.

1   AZZO, J. D. and HOUPIS, C. H.: 'Feedback control system analysis and synthesis', (Mc-Graw Hill, 1966)

2   KUO, B. C.: 'Automatic control systems', (Prentice-Hall Inc., 1966)

3   NETUSIL, A.: 'Theory of automatic control', (MIR Publishers, Moscow, 1973)

4   ÅSTRÖM, K. J. and WITTENMARK, B.: 'Computer controlled systems: theory and design', (Prentice-Hall Inc., New Jersey, 1990)

5   ISERMANN, R.: 'Digital control systems', (Springer Verlag, Berlin, 1989)

6   MIDDLETON, R. H. and GOODWIN, G. C.: 'Digital control and estimation: a unified approach', (Prentice-Hall Inc., New Jersey, 1990)

Chapter 4

Integral wind-up in control and system simulation

B. Šulc

In this chapter a problem of practical controller implementation that is often disregarded is presented, with the aim of demonstrating reasons for its occurrence, consequences if it is neglected, and some technical solutions for its removal. Close links to a correct control system simulation will be also shown.

4.1 Introduction

In each technical implementation of a control system the magnitude of the manipulated variable is limited. In the cases of large changes in the controlled variable, the controller output (manipulated variable) exceeds the values that can be set by an actuator and the manipulated variable becomes saturated. This non-linear effect leads to different control responses in comparison with those obtained from a linear control system model. This non-linearity is not the only reason for this difference. Actuator saturation is very often connected with existence of so-called 'reset wind-up'. Reset wind-up is a phenomenon restricted to controllers providing integral action. Some of these controllers do not give any guarantee that their integral action parts will stop the integration operation as soon as the controller output exceeds limiting values. Continuing integration produces excess values that can never be achieved by the actuator. However, in the control response the period when the controller output is limited (manipulated variable is saturated) appears as a temporary loss of feedback or a temporary time delay in control action. Depending on how large the integral excess becomes, the control response is characterised by a slower response, sometimes oscillating, with a large overshoot; in the worst case an instability phenomenon can occur. For these reasons reset anti-wind-up measures represent one of the most important components in the technical realisation of digital controllers. Integral action in digital controllers is carried out numerically and therefore there is no other physically-based reason for an integral action stopping but the limits of number representation in the computer. This is why digital controllers may have very different behaviour in actual application compared to that expected from the linear theory. Simulation of such linear control systems should respect all true magnitude bounds in the plant models and take into account necessary integration stopping instead of the integration cropping usually applied.

4.2   Motivation and control problem statement

Integration performed by hardware means it is always going to stop as soon as the integrator output reaches one of its limits. These limits depend on technical principles which are used in the instrumentation available for performing integration. However, it happens (especially often in controllers or simulation models) that the output of the system, perhaps due to physical or technical restriction, has a smaller range of output values compared with the internal values of the outputs during integration. In such cases the output saturates and integral (reset) wind-up may occur.

The integral wind-up is manifested by maintaining changes in the internal integrator outputs (provided their physical or technical limits allow) after the output of the system under consideration has already reached one of its limiting values. The continuing integration leads to an excessive nonrealisable output. Removing the integrator output value excess and returning its values into a technically or physically acceptable range is possible only after a corresponding input change (causing an opposite trend in integration). However, such removal takes time and this is (like a transport time delay) a reason for differences in behaviour between real control circuits and their linear models. This is an important problem concerning both correct process simulation and proper controller function.

The goals in considering this problem are as follows:

— to show the occurrence of reset wind-up in connection with different constructions of PI controllers;
— to refer to certain approaches to anti-wind-up techniques;
— to demonstrate the importance of a correct control process simulation with respect to the integral wind-up effects appearing both in various constructions of controllers and in simulated plant models;
— to present experiments for a laboratory rig which allow practical measurement of control responses with and without wind-up.

4.3   Technical aspect of integral wind-up

The problem of integral wind-up can be explained by analysing its occurrence in different constructions of the PI controller. It is usually referred to reset wind-up if integral action of PI or PID controllers is intended.

Inherent differences in behaviour between controllers caused by reset wind-up can be demonstrated by the step responses in Figure 4.1. The maximum realisable output value is considered to be 100%. For the controller A with zero wind-up this value cannot be exceeded, because the integration will come to a standstill as soon as this limit value of the output is reached. The case of a controller B with limited wind-up is characterised by continuing integration (after reaching the limit value of 100%) but not exceeding a technical maximum (e.g. 135%). The full reset wind-up X has no limits for stopping integration. A control error sign change need not be the sufficient reason for return to the region of effective operation.

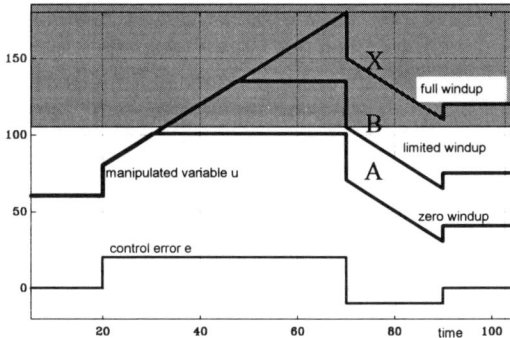

Figure 4.1   Step responses of PI controllers with different levels of reset wind-up

4.3.1   Reset wind-up occurrence in PI controllers of different construction

The significance of reset wind-up varies according to the construction of the controller. There are controllers in which this problem does not occur at all (e.g. in the hydraulic PI controller hereinafter presented). On the other hand, in the digital form of a PI controller, special attention must be paid to avoid a very strong reset wind-up effect.

Hydraulic PI controller
A simplified scheme for a hydraulic realisation of the PI controller is presented in Figure 4.2. In this construction the integral action is performed by the piston

movement, the speed of which is proportional to the flow rate from an oil supply. The flow rate is again proportional to the control error. This becomes zero if the piston cannot move any further at the end (limit) positions, or if the control error is zero. Integration is automatically stopped at the end positions and thus no wind-up effect can occur (see Figure 4.1A). As soon as the control error changes its sign, the output immediately leaves the limit value it has just reached, but in the opposite direction.

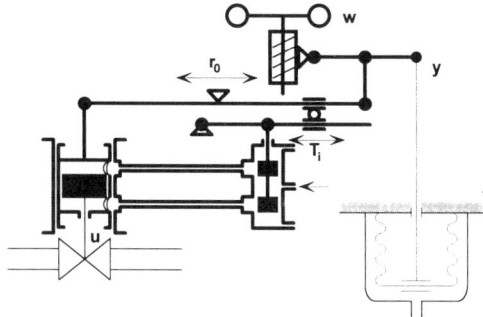

Figure 4.2   Hydraulic PI controller with zero wind-up

Pneumatic PI controller
In most pneumatic PI controllers pneumatic capacity is used for producing integral action. Extreme values of pressure in such capacity (e.g. the pressure P_I in the integrating bellows in one of the possible technical solutions depicted in Figure 4.3) cannot exceed the supply maximum (usually $P_N = 130$–$140\,kPa$) or a minimum ($0\,kPa$). These values limit the wind-up effect. That is, the output of such a pneumatic controller cannot exceed approximately 40% of the standardised output signal range 20–$100\,kPa$. Since only the output values within the standardised range can be set by actuators, control error changes do not have any influence on the manipulated variable until the pressure in the integrating capacity is brought back into the working range by the integration (see Figure 4.1B). Due to the simultaneous operation of the proportional action, the time when the controller does not respond to the control error change is usually short and its influence on control behaviour can often be disregarded.

Digital PI controller
A block diagram of a digital form of the PI controller is illustrated in Figure 4.4. The incrementation of the output u_k, which always appears in the case of a non-zero control error, is practically unlimited. The output increases or decreases as long as the control error does not change its sign and the numerical range is not exceeded.

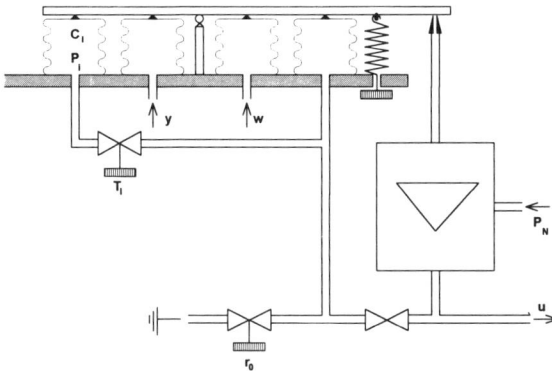

Figure 4.3   Pneumatic PI controller with limited wind-up

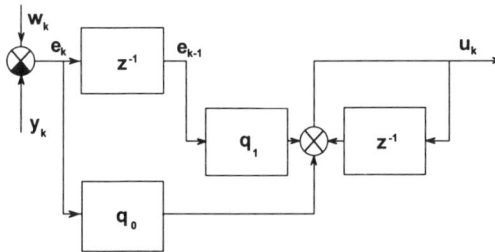

Figure 4.4   Block diagram of a digital PI controller with full wind-up

4.3.2   Anti-wind-up measures

Anti-wind-up measures are usually based on two main principles: integral operation stopping, or integral operation affecting so that the undesirable integrator output excess can be excluded (or at least reduced). Both require information on the actual values of the manipulated variable set by the actuator, but for the integration stop 'on/off' information from end switches is sufficient. Information is obtained either by measurement of real manipulated variable values or by modelling magnitude and rate limitations of the actuator.

From the techniques developed for dealing with the problem of reset wind-up the following methods may be mentioned:

— fixed limits on integral term
— stop integration (conditional integration)
— velocity (incremental) algorithms
— integral subtraction (back calculation; tracking) [1]

— conditioning technique [2]
— observer approach [3,4]
— analytical method [5]

4.3.3   Bumpless transfer

Controllers for industrial applications generally require the possibility to switch between automatic control and manual control. There are two conditions for efficient transfer: first, a good tracking performance is required after switching to manual and secondly, the transfer in both directions, but especially back from manual to automatic, must be carried out without any bump at the instant of switching. Bumpless transfer from manual to automatic mode will occur providing the output of the disconnected controller tracks exactly the actual value of the manually changed manipulated variable. This can be achieved very easily by incremental algorithms (see Figure 4.5). The conditioning technique also offers bumpless transfer, which is referred to as conditioned transfer. In tests by Vrancic [6], conditioned transfer provided faster return to the reference value than the incremental technique.

Controllers without reset anti-wind-up respond to the control error with an unlimited increase or decrease of their output when disconnected during manual control. At the moment of re-connection, great disparity exists between the value of the manipulated variable produced by the controller and the value set manually which causes a transitional 'bump'. The rate of recovery from this 'disturbance' is related to the dynamics of the process. Overall this approach should be considered a poor technical solution.

4.3.4   Incremental PID algorithm for practical applications

In Figure 4.5 a block diagram is shown for an incremental PID algorithm consisting of four (dotted) blocks which perform the additional functions necessary for a digital controller in practical applications. The three-term PID algorithm is performed in block 3, which also involves transformation of continuous PID controller parameters into those of the discrete form. Inputting of r_0, T_I, T_D instead of q_0, q_1, q_2 is often preferred in controller adjustment. Block 4 contains a limiter as a model for actuator saturation. The output value u_k cannot exceed both limit values u_{min} and u_{max}. If some of these limit values have been reached then the control u_k immediately leaves them as soon as the sign of the increment Δu_k allows it. This protects the controller against reset wind-up. In blocks 1 and 2 an accelerated incremental setting (the time increment is $1/r$ of the sampling period) of the set point w_k (if switched to mode A, 'automatic') or the manipulated variable u_k (if switched to mode M, 'manual') is used. No settings beyond the operating area are possible. Block 1 ensures that the set point has the same value w_k as the controlled variable y_k during manual operation. No bump will occur at the moment of changeover to automatic mode as the control error is zero throughout manual operation. If there are no other reasons (such as disturbances, transient state), no control actions will be undertaken after transfer from manual to automatic mode. This set point tracking seems to be a reasonable tool when rapid transient responses are expected when switching from manual control.

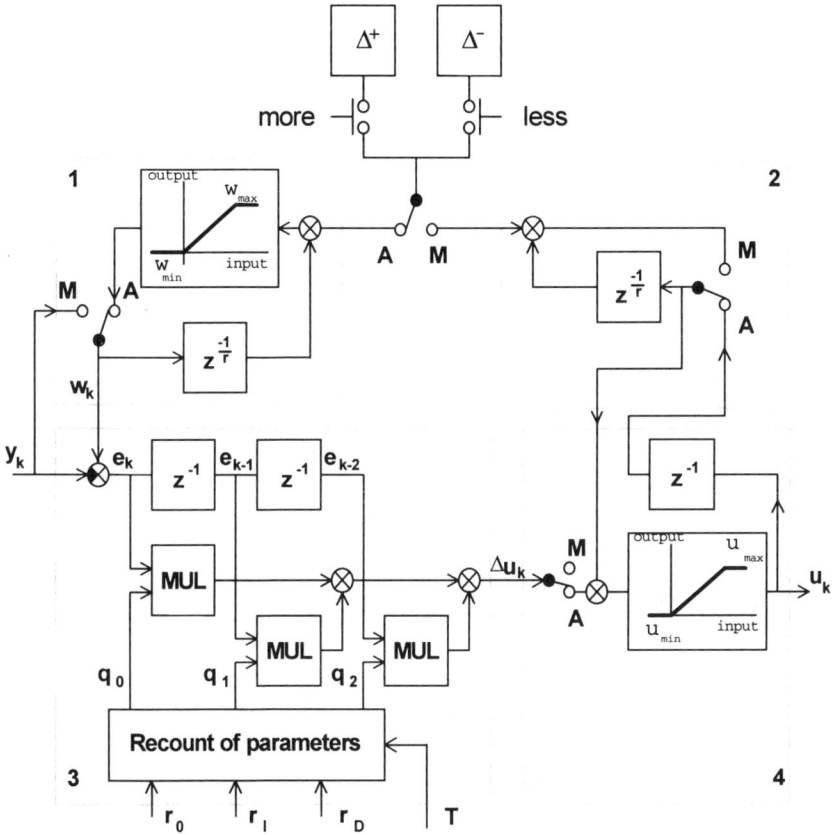

Figure 4.5   *Incremental digital PID controller (3) with reset anti-wind-up (4), bumpless transfer (2) and incremental control command setting (1)*

4.4   Discussion of anti-wind-up approaches

Several comparisons of anti-wind-up techniques have recently been made [7]. They are usually based on linear simulation models and this can be dangerous for quantitative interpretation of their results. Essentially, real occurrences of a wind-up are connected with large changes of the set point or load and depend on the starting position of the operating point and controller setting. For this reason the comparisons are more or less specific to a particular simulated situation. In well designed real control loops it can be difficult to produce wind-up behaviour. From a practical viewpoint, only those anti-wind-up solutions that provide a bumpless switching facility are useful. Although often applied, actuator saturation modelling is not always recommended [8], and methods satisfied by simple hardware measurement (end micro-switch) should be preferred (e.g. the first four in the summary list in Section 4.3.4). Most of the approaches quoted have been derived using a continuous linear description of plant, which can restrict their use in discrete time for certain sampling period selections.

The incremental ID algorithm presented (see Section 4.4.4) satisfies these requirements. However, analysing terms in the difference equation for a discrete PI controller (and for PID control this reasoning is even more comprehensive):

$$\Delta u_k = r_0 \Delta e_k + \frac{r_0}{T_I} Te_{k-1}$$

it can easily be recognised that the increment Δu_k becomes negative even when the control error remains positive. A large negative increment Δe_k (which often appears during the first phase of response) can cause the first term (which is in this case negative) to prevail over the second positive one and the last value u_{k-1} of the manipulated variable, which has reached the upper limit value of saturation, begins to decrease although the longer lasting positive control error requires it to remain at maximum. Such behaviour may be affected by the setting of the controller parameters, but does not occur in the conditioning technique approach [6].

4.5   Laboratory set-up and simulation model for wind-up investigation

Given that the occurrence of the wind-up phenomenon in real conditions is connected with large changes in load or operating point, the best way to obtain objective conclusions on wind-up would be testing on a real device. This is rarely possible under industrial conditions, but laboratory set-up models of the most common kind of real control loops, whose generalised scheme is shown in Figure

4.6, can provide us with valuable experience. At the same time, the problem of correct simulation using linear models for such real control loop simulation can easily be demonstrated for the considered class of first-order plants.

4.5.1 Integration stopping in simulation models

Simulation is a very efficient tool for a graphical demonstration of integral wind-up consequences. There are two different situations in simulation where the integral wind-up phenomenon needs to be taken into account. In the first, only a comparison of control loop behaviour with or without reset wind-up is required. A model of a controller providing both abilities is needed. One possible solution for integration stopping is depicted in Figure 4.7 for a continuous PI controller, where the block INT represents an integration block, MUL is a block performing multiplication of all its inputs and the blocks OR provide a logical operation "OR" with numerical results 0 or 1. Multiplication by a constant is made by blocks bearing symbols for the controller gain r_0 and the integral time constant T_I. The remaining blocks represent a transformation from an analogue signal to a binary one (in order to produce numerical values 0 and 1).

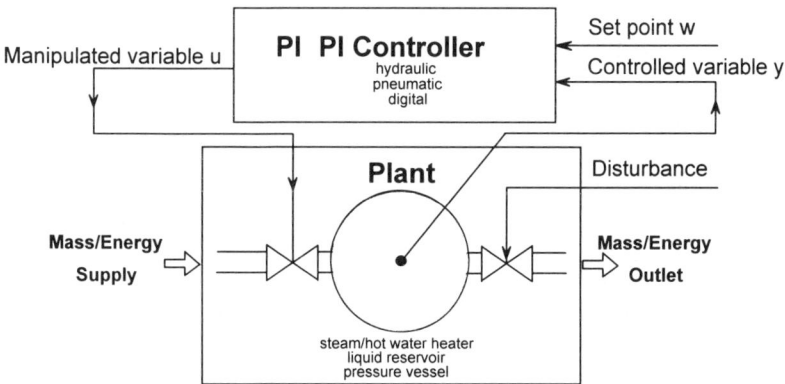

Figure 4.6   *Generalised control loop scheme for PI control of plants behaving as a first-order system*

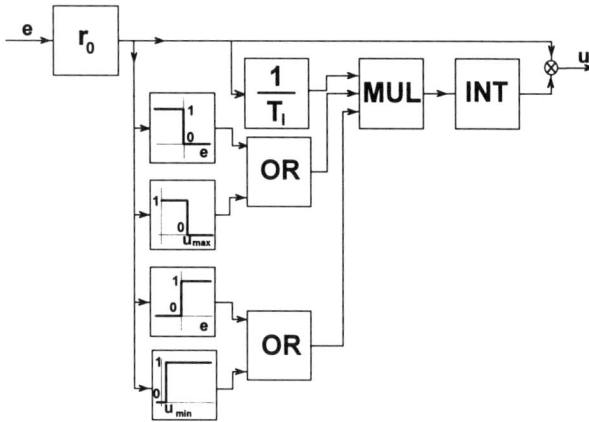

Figure 4.7   Integration stopping in a PI controller simulation model

The second situation is less well known but very realistic, and should be considered in any simulation where there is a danger that the variables may reach values that do not correspond to physical or technical conditions. In quoted examples of plants with one mass/energy accumulator, which are modelled by linear first-order systems, it is possible that the controlled variable y (the output from the linear model) can exceed the real value range determined by technical or physical limits. Such non-exceedible maxima are 100 °C in water heaters, a maximum level of a liquid in a tank or a maximum pressure in a vessel etc. This fact should be borne in mind in simulation models. The same anti-wind-up precaution must then be included in the integrators of the simulated plant model if a correct simulation of realistic control loop behaviour is required. Integrators of the simulated plant model must be equipped with an integration stop which starts to work if changes of the output variable bring the plant model output beyond the technically or physically realisable range. The solution most often used in simulation, which applies a limiter in order to restrict the output to a permitted range, is not a valid approach.

This is demonstrated by Figure 4.8 which depicts the step responses of a linear first-order model of a heater. If the step change of the manipulated variable u is large enough and no integration stop is used, then the temperature either exceeds 100 °C or, if it is limited in its values by a limiter, it will respond with a 'wind-up' delay when leaving this limit value after the input returns back to the starting value. This delay does not occur in reality; it is a consequence of an integral wind-up in simulation. Supposing there are overshoots over such physical limits during the control process simulation, then all the integrators in the simulation model of a plant must be treated against wind-up.

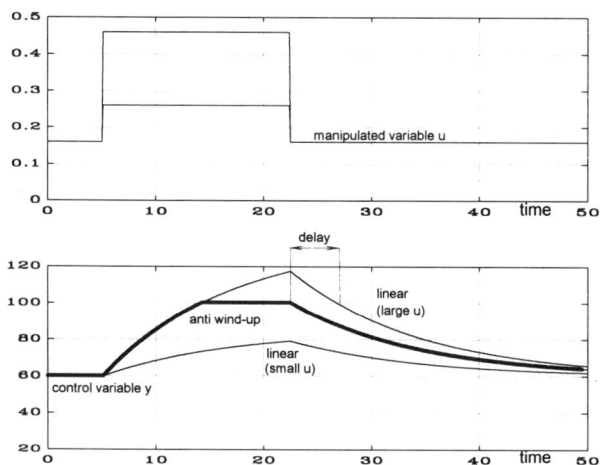

Figure 4.8   Integral wind-up in plant simulation

4.5.2   Laboratory set-up description

Experiments suggested in the following section can be carried out on every physical model of a control circuit which offers digital control facilities programmable for both alternatives of PI or PID control algorithms (i.e. with and without anti-wind-up precautions). The results presented here were obtained from the set-up depicted in Figure 4.9.

It consists of (1) a pressure vessel with the controlled variable y representing pressure inside the vessel; (2) a microcomputer by which the PI parameters (r_0, T_I) are defined, the sampling period T is set and values w of the set point input; (3) a control valve; and (4) an adjustable pressure source. The rig includes necessary transmitters and a line recorder for easier evaluation of results. The control programme (incremental PI controller) is saved in EPROM memory. Anti-wind-up measures can be switched on or off by pressing a key.

4.6   Suggested experiments

Suggested experiments might be divided into two groups: experiments performed by means of the set-up and simulation experiments. Modern simulation tools allow experiments to be performed in a way very similar to set-up experiments. Therefore there is no need to distinguish the two groups. The following are suggested:

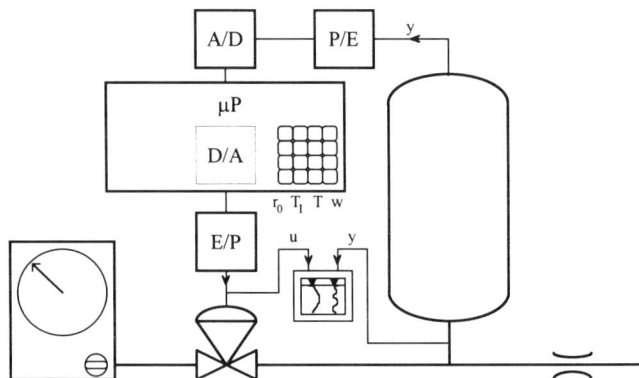

Figure 4.9   Laboratory set-up (control of pressure in a vessel)

— to test the controller function especially to find out whether the algorithm
    is set to provide reset anti-wind-up or not (in the set-up described,
    exclusion of the plant dynamics can be achieved by closing the outlet from
    the vessel and using adjustable pressure resource for manual generation of
    the controlled variable y)
— to determine conditions under which wind-up appears and to verify the
    efficacy of reset anti-wind-up measures
— to compare simulation results with real measurements, especially from the
    viewpoint of correct simulation as demonstrated by the illustrative results
    in the next section.

4.7   Illustrative results

4.7.1   Results from set-up measurement

Several records from the line recorder are shown in Figure 4.10 so that an easy
comparison of control results can be made. The responses of the control variable y
are caused by changes of the set point from 22 to 90 kPa and back. The
experiments when the anti wind-up measures in the control algorithm were
switched off are depicted with dotted lines. Some non-linear effects in the E/P and
P/E transmitter, and a large change in the plant gain, have led to differences in
comparison with theoretically expected results. The controller setting was not
changed during experiments.

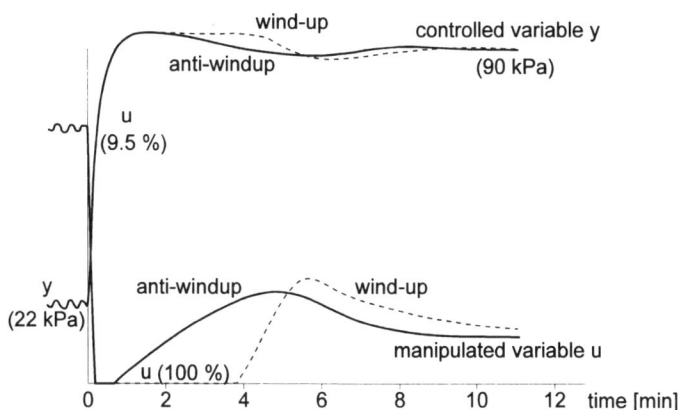

Figure 4.10   Set-point responses from experiments

4.7.2   Results from simulation

The differences that appear in simulation of the control loop behaviour according to a different consideration of wind-up effects are shown in Figure 4.11. Simulation models of the PI controller and the first-order plant are linear in all the presented cases, but only the case of a LIN (fully linear control loop model) does not involve any restriction of a non-linear character. Therefore the manipulated variable u enormously exceeds the operating range 0–1 in this case, and therefore the return to the original set point value 60 °C is connected with a very unrealistic response (negative values of u mean 'cooling'!). The traces of integral action extracted from the responses of the manipulated variable u in the middle graphs of Figure 4.11 explain the reasons for the great difference between control using reset anti-wind-up precautions (AWP) and all other cases which prefer limitation of the magnitude of the manipulated variable to the reset anti-wind-up precaution (PIS, LIS). The integral action in the manipulated variable u in the case of AWP will not change further once values of the controller output u exceed the operating range, and therefore the integral action can never exceed the values 0 or 1. In contrast, in the cases of LIS and PIS, the integral action is allowed to reach any possible value outside the limits 0 or 1. The LIS case (linear control loop with actuator saturation) demonstrates what happens if no integral anti-wind-up is used in the simulation model (both in the controller and plant models), but the usual restriction on the magnitude of the manipulated variable is applied. In this case the overshoot reaches an unrealistic 115 °C. The plant integration stop applied in the PIS case leads to a much more realistic simulation and to a better control response. This contribution can be seen only in the case of a positive set point change, where integral anti-wind-up really operates. A small excess over 100 °C is caused by the numerical computation of the integral and by the stopping algorithm used in which the last value after crossing limits represents the integral output for the whole time it is stopped.

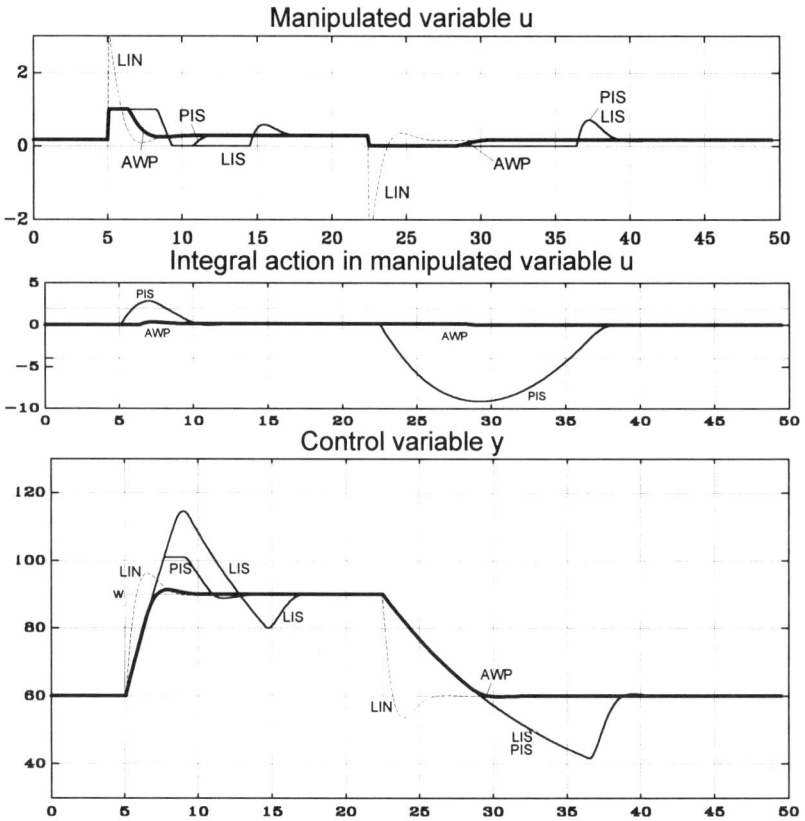

Figure 4.11   Control loop simulation with various level of the integral anti-wind-up precautions

AWP	full anti-wind-up (both controller and plant integrator stop)	LIN	purely linear PI control (no restrictions for u and y)
PIS	plant integrator stop (only for plant integrator anti-wind-up)	LIS	linear PI control with actuator saturation (full wind-up)

4.8 Conclusions

Integral wind-up is a phenomenon whose negative consequences are evident both in control and simulation. These can be removed or suppressed in a relatively straightforward way enabling significant improvement of control behaviour, especially for large changes in load or in set point setting. Bumpless transfer of a high efficiency can also be ensured providing application of reset anti-wind-up techniques. Not only should algorithms of PID type be equipped with reset anti-wind-up precautions but also more sophisticated algorithms with an optimised structure (e.g. dead-beat, pole assignment, quadratic optimal etc.) should be designed to take into account actuator saturation if they are intended for real practical applications. The experiments suggested can help in familiarisation with this topic that is of great importance for digital control.

4.9 References

1   ÅSTRÖM, K. J. and RUNDQWIST, L.: 'Integrator wind-up and how to avoid it', American Control Conference, 1989, **2**, pp. 1693–1698

2   HANUS, R.: 'A new technique for preventing Wind-up', *Journal A*, 1980, **21**, (1), pp. 15–20

3   BENNETT, S.: 'Real-time computer control: an introduction', (Prentice-Hall Inc., 1988)

4   BENNETT, S. and LINKENS, D. A. (Eds.): 'Real-time computer control', (Peter Peregrinus Ltd. on behalf of the IEE, 1984)

5   CAMPO, D. J. and MORARI, M.: 'Robust control of processes subject to saturation nonlinearities', *Computers Chem. Engng.,* 1990, **14**, (4/5), pp. 343–388

6   VRANCIC, D. and PENG, Y.: 'Tutorial: anti-wind-up, bumpless and conditioned transfer techniques for PID controller', Report DP-6947, 1994, J. Stefan Institute, Ljubljana, Slovenia and Dept. Control Engineering, Free University of Brussels

4.10 Other reading

ÅSTRÖM, K. J. and WITTENMARK, B.: 'Computer control systems: theory and design', (Prentice-Hall Inc., 1984) (Russian translation: *MIR*, Moscow, 1987, 2nd Edn., 1990)

HANUS, R., KINNAERT, M. and HENROTTE, J. L.: 'Conditioning technique, a general anti-wind-up and bumpless transfer method', *Automatica*, 1987, **23**, (6), pp. 729–739

HANUS, R.: 'Anti-wind-up and bumpless transfer: a survey', *in* BORNE, P. *et al.* (Eds.) 'Computers for control systems', IMACS, 1989, **4**, pp. 3–9

ISERMANN, R.: 'Digital control systems, Vol.1: fundamentals and deterministic control', (Springer Verlag, Berlin, 1989)

MORARI, M.: 'Some control problems in the process industries. Essay on control: perspectives in the theory and applications', (Birkhäuser, 1993) pp. 56–77

NOISSER, R.: 'Anti-reset-wind-up-Maßnahmen für Eingrößenregelungen mit digitalen Reglern', *Automatisierungstechnik*, 1987, **35**, (12), pp.499–504 (in German)

ŠULC, B.: 'Dead-beat control system design with actuator saturation' *in* 'Advanced approaches in industrial control systems', TEMPUS Workshop Proceedings, Prague 1993, SEFI Document 8/1993, pp. 113-116

TAKAHASHI, Y., RABINS, M. J. and AUSLANDER, D. M.: 'Control and dynamic systems', (Addison-Wesley Publishing Co. Inc., London, 1972)

THOMAS, H. W., SANDOZ, D. J. and THOMPSON, M.: 'New de-saturation strategy for digital PID controllers', *IEE Proc. D*, 1993, **130**, (4), pp. 188–191

WHITWORTH, C.: 'The survey and evaluation of various anti-wind-up techniques for the digital PID controller', B.Eng. Report, 1994, Dept. Automatic Control and System Engineering, University of Sheffield

WITTENMARK, B.: 'Integrators, nonlinearities and anti-reset windup for different control structures', American Control Conference, 1989, **2**, pp. 1679–1683

Chapter 5

Control of unstable systems

Đ. Juricic and J. Kocijan

5.1   Introduction

Many practitioners and researchers in the field of control agree that control systems design can be quite laborious in cases where the plant is unstable. What makes the controller synthesis of unstable systems particularly interesting is the fact that there are, objectively, certain design and closed-loop performance limitations that narrow the range of feasible solutions. These constraints reflect in overshoots, peaks in sensitivity functions and closed-loop bandwidth. Choice of the control structure and the design approach depends considerably on *a priori* knowledge of the process dynamics and system requirements. The aim of this chapter is to point out some problems occurring in unstable systems control and review some of the basic options for stabilisation of such systems. Finally, their properties are demonstrated on a laboratory test plant and discussed from several perspectives relevant to practitioners.

5.2   Motivation (control problem)

Unstable open loop behaviour is a property inherent to a number of systems. Pure unstable modes usually occur in flight systems [1,2], whereas in process industries they are relatively infrequent. As an example, let us mention exothermic reactors [3] where heat (produced by chemical reaction) increases the temperature which, in turn, increases the rate of reaction, and hence the entire system becomes unstable.

Control design of unstable systems could be attempted using many of the methods applicable to ordinary stable systems. However, instabilities imply some constraints which require careful and sensible design. Two classes of constraints can be expected in cases of unstable systems:

— design limitations
— performance limitations.

The main purpose of this chapter is to provide a clear summary of these limitations and to demonstrate some of the prospective techniques on a laboratory test plant.

5.2.1   Design limitations

System identification
A good process model is a standard prerequisite for good control design. If physical modelling is not feasible or reasonable, identification approaches must be used. However, direct open-loop estimation of unstable systems is not recommended due to numerical stability, as well as problems with bias in estimates [4]. Instead, it is better first to stabilise the plant by means of a controller with a known transfer function and then estimate the closed-loop parameters using some of the well known techniques suited to stable systems [5,6]. Recalculation of the original plant parameters can be made in a straightforward manner.

Model validation
How good is a model of an unstable process? Unfortunately, it is not easy to 'measure' the quality of unstable process models using most of the ordinary metrics applicable to the stable plants. Namely, any inconsistency in the model implies unbounded increase of process-model mismatch so that metrics like mean-squared error turn out to be useless.

Global stabilisability under constrained input
Provided the input to an unstable system is constrained within an interval $U=[u_{min}, u_{max}]$ there exists a constrained region (region of stabilisability) in the state space in which the system can be stabilised by using control actions $u(t)$ from the set U. If, by accident, the system state falls out of the region of stabilisability, then there is no feedback law that can stabilise the plant provided $u(t) \in U$ [7]. This has to be considered in the early phase of system design.

Cautious controller design
Consider a closed-loop system as shown in Figure 5.1. Let the return ratio $H(s) = C(s)G(s)$ contain real right half poles s_i ($i = 1,...,k$). Then the integral of the sensitivity function $S(s) = (1+H(s))^{-1}$ obeys the following law [8, 9]:

$$\int_0^\infty \log|S(jw)|\ dw\ =\ \pi \sum_{i=1}^{k} s_i \qquad (5.1)$$

Since much of the positive area must lie below the crossover frequency [8], it follows that reduction of sensitivity at lower frequencies will result in higher peak values at higher frequencies. This renders the appropriate controller design considerably more difficult than in the case of stable open-loop poles when the right side of Equation 5.1 is zero.

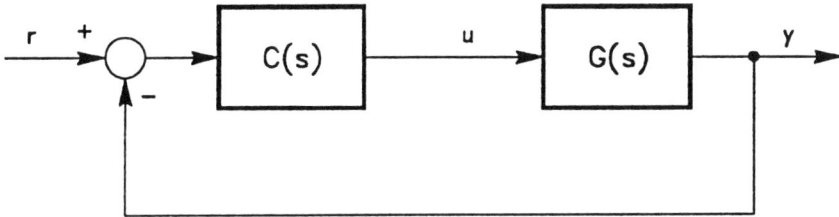

Figure 5.1   *Basic unity feedback system*

5.2.2   *Performance limitations*

If $H(s)$ contains right half poles and if the closed-loop system $H(s)/\{1+H(s)\}$ is stable, then overshoot might be expected in the closed-loop step response [10]. The amount of overshoot is closely related to the rise time. Surprisingly, the longer the rise time the higher are the overshoots, which is contrary to the case of stable plants.

Furthermore, right half plane poles dictate the lower boundary of the closed loop bandwidth ω_B in the following manner [10]:

$$\omega_B \geq 5 \sum_{i=1}^{k} s_i \tag{5.2}$$

This result means that unstable poles impose the need for fast action on the control input. However, this must be considered very carefully when noise acts in measured output.

5.2.3   *Control problem*

The stimulus for the following study originates from diving. Imagine a diver who has to reach a certain depth, perform a task and return to the surface. On the way he/she has to stop at prescribed depths to perform decompressions within pre-specified retention times. A simplified sketch of the problem is given in Figure 5.2. Movement of the diver is regulated by means of the volume of the attached balloon which is filled (and emptied) with air from the air supply. The idea is to provide a stand-by automatic controller which could take action in case of emergency and hence safely bring the diver to the surface by proper blowing and emptying of the balloon, thus leading him through the required decompression phases.

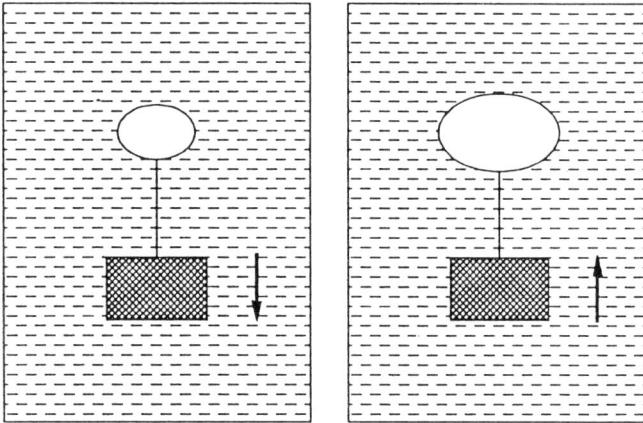

Figure 5.2   Principle of diving

To study interesting control problems a laboratory test plant has been constructed as shown in Figure 5.3. The plant consists of a tube and a reservoir linked by a pipe. The tube is filled with water but so that an air cushion is left at the top. The very top of the tube is hermetically closed. The dynamic part of the plant is a body ('rocket') floating up and down on an air bubble within the tube (akin to our diver!). The vertical position of the object depends on the size of the bubble which in turn depends on the hydrostatic pressure of the water. The higher the water pressure acting on the bubble, the smaller is its volume. Consequently, the weight of the rocket becomes greater than the buoyancy force and it falls. As the water pressure decreases, the bubble increases in volume and buoyancy force pulls the rocket up. If both forces are equal, the rocket remains (theoretically) motionless. This state is, however, difficult to maintain since there is no stable equilibrium point. The air pressure at the top of the tube is influenced by the voltage applied to the pump. The control problem is to keep the rocket at the reference position by manipulating the pump voltage.

Problems which must be taken into account include the following:

— in ordinary operating conditions, only the position of the rocket
  is measured;
— the pump's static characteristic is nonlinear;
— the dynamics of the rocket are unstable.

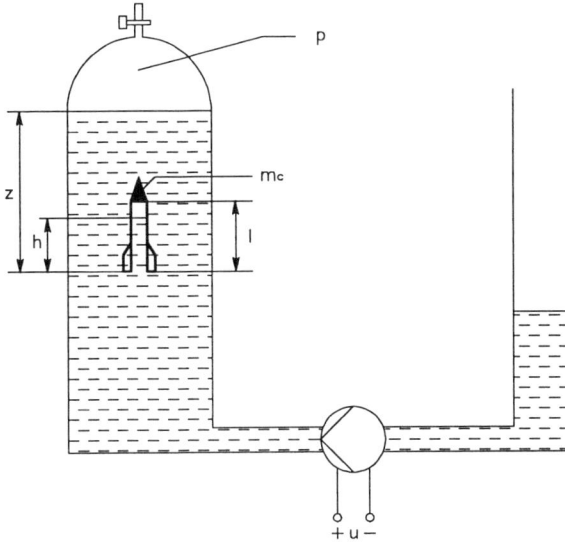

Figure 5.3   Laboratory test plant

Mathematical modelling

The acceleration of the rocket is governed by the difference between its own weight and the buoyancy force:

$$m_c \ddot{z} = m_c g \left(1 - \frac{\rho}{m_c} \left(V_r + S(l - h) \right) \right) \tag{5.3}$$

where m_c, S, l, and V_r represent mass, cross-sectional area, length and volume of the rocket, respectively. These values are: $m_c = 0.09075$ kg, $S = 3.8 \times 10^{-4}$ m^2, $V_r = 95.85 \times 10^{-6}$ m^3 and $l = 0.145$ m. The water level h (see Figure 5.3) is a function of pressure p within the cushion and rocket position z:

$$h(z, p) = \frac{1}{2} \left(\frac{p}{\rho g} + z + l - \sqrt{\left(\frac{p}{\rho g} + z + l \right)^2 - 4 \left(\frac{(p - p_0)l}{\rho g} + zl \right)} \right) \tag{5.4}$$

where p_0 is the normal air pressure. The air pressure within the cushion depends on the pump voltage u as follows:

$$\dot{p} = a\left\{-p + \alpha + \beta(t) + \gamma u^2(t)\right\} \tag{5.5}$$

where a, α, β and γ are appropriate constants with the values $a = 2$, $\alpha = 95645$, $\beta = 401.5$ and $\gamma = 476$. A linearised model of the entire plant, calculated at the stationary point defined by $u^* = 4.3\,\text{V}$, in the state space form is as follows:

$$\dot{x} = Ax + Bu$$
$$y = Cx \tag{5.6}$$

where

$$A = \begin{bmatrix} 0 & 1 & 0 \\ 0.4755 & 0 & 4.845E-5 \\ 0 & 0 & -2 \end{bmatrix}, \ B = \begin{bmatrix} 0 \\ 0 \\ 9E3 \end{bmatrix}, \ C = \begin{bmatrix} 1 & 0 & 0 \end{bmatrix} \tag{5.7}$$

and state vector $x = [z, dz/dt, p]'$. The system poles are located at ± 0.6896 and -2.

5.3   Technical approaches to the control of unstable processes

Choice of the controller structure, as well as adjustment of structure parameters, depends on several factors [11–13]:

— the amount of knowledge about the plant;
— the performance requirements;
— the possibilities for and cost of experimentation;
— the design costs

A schematic representation is given in Table 5.1. In the following only techniques marked with '✓' will be treated.

		Controller structure		
		State controller	State controller with observer	PID
Design	Model based	✓	✓	✓
	Model-free (GA approach)			✓

Table 5.1   Some of the approaches to unstable systems control

5.3.1 State controller

Optimal state control is a well-established topic in control theory and a number of textbooks and papers address the problem (e.g. References 14–17).

To summarise, let us take a controllable plant:

$$\frac{d}{dt}x = Ax + Bu$$

$$y = Cx$$

(5.8)

with $x \in R^n$, $u \in R^m$, $y \in R^p$, $A \in R^{n \times n}$, $B \in R^{n \times m}$, $C \in R^{p \times n}$. If $n = p$ the optimal control law is then of the form:

$$u = -Kx$$

(5.9)

A way to obtain the matrix K is to minimise the quadratic cost function:

$$J = \int_0^\infty \left(x(t)'Qx(t) + u(t)'Ru(t) \right) dt$$

(5.10)

with Q and R being symmetric and $Q \geq 0$ are $R > 0$. The result is:

$$K = R^{-1}B'P$$

(5.11)

where P is a solution of the matrix Riccati equation:

$$PA + A'P - PBR^{-1}B'P + Q = 0$$

(5.12)

Such a controller is called a linear quadratic regulator (LQR). Another alternative to the derivation of the state controller is pole placement where K is such that $\det(A-BK)$ has zeros at the prescribed locations [14].

5.3.2 State controller with observer

It is seldom the case that all system states are available. Unmeasurable states can be reconstructed from measurable outputs by means of the state observer. The full-order observer for the system in Equation 5.8 is of the form

$$\frac{d}{dt}\hat{x} = A\hat{x}(t) + Bu(t) + L(y(t) - C\hat{x}(t))$$

(5.13)

where $\hat{x} \in R^n$ is the estimated state and $L \in R^{nxp}$ is the observer gain. In that case the optimal control law is

$$u = -K\hat{x}$$ (5.14)

Matrix L must be such that all eigenvalues of the matrix $(A–LC)$ lie in the left half of the complex plane. A necessary and sufficient condition for asymptotic stability of the observer is that the pair (A,C) is detectable [15]. The observer design is dual to that of the state regulator, i.e. the dual to the LQR is the Kalman filter [15].

5.3.3   PID controller

A transfer function for a real PID controller is as follows:

$$C(s) = K_p\left[1 + \frac{1}{T_i s} + \frac{T_d s}{T_s s + 1}\right]$$ (5.15)

where T_s is normally assumed to be sufficiently small. A plethora of different approaches to PID tuning exist (see, for example, [18–20]) which are applicable to stable processes or processes with one integrator pole. These approaches are based on simplified process models but, unfortunately, they are of no use in cases of unstable systems. If a model-based design is used, then a more elaborate process model is required. Design could be easily performed, for example, in the frequency space by means of a Bode diagram [21].

5.3.4   Model-free design of controllers — a genetic algorithm approach

If the unstable process model is not known, the controller parameters should be determined by experimentation, i.e. by means of on-line optimisation. A particularly effective technique appears to be that of genetic algorithms (GAs) which are but a class of adaptive and general-purpose learning algorithms based on the laws of natural genetics. The effectiveness of GAs is achieved by propagating the best partial solutions through the search [22–24]. GAs are well suited for non-convex criterion functions where additional restrictions to the optimum could be imposed [25–28].

5.4   Discussion

Each of the controllers suggested has both advantages and drawbacks.

5.4.1 Linear quadratic regulator

The main advantage of the LQR is the fact that at high frequencies it behaves like a first-order system. At the 0 dB cross-over frequency the phase margin is at least 60° whilst above that frequency the phase curve gets close to –90°. Hence the LQR has an infinite gain margin in the classical sense [9,15].

Another advantage of the LQR is that it is possible to deliberately locate all the plant poles in given regions of the complex plane. However, in virtually all cases of linear quadratic design (minimisation of criterion 5.10) there is a region in the complex plane where it is not possible to assign the closed-loop poles by classical combination of weights Q and R [29]. In the same paper the author has proposed a remedy to avoid 'hidden' regions, the result of which is that (surprisingly) some of the weights in Q could become negative!

The disadvantage of full-state feedback is that all the states must be measurable, which is seldom the case.

5.4.2 Linear quadratic regulator with observer

For a reduced set of measured variables a state observer must be added to the state controller. In principle, the state space controller and observer can be designed separately since each of them exhibits reasonable properties of robustness and performance. Unfortunately, when put together they may lead to degraded robustness and performance characteristics [30,31]. However, there are a number of techniques which lead to robust design [9,32,33]. For example, in loop transfer recovery (essentially) some of the filter's eigenvalues are located at the zeros of the plant whilst the remaining poles are allowed to be arbitrarily fast [9].

5.4.3 PID controller

The PID controller has a simple structure (only 3 tuning knobs) so that it can be tuned relatively easy even 'by hand' (though in the case of unstable plants this is not trivial). If a model-based design is used, it is clear that in systems of an order higher than two some poles could be not freely assigned.

5.4.4 Model-free design using a genetic algorithm approach

In principle any control structure could be tuned using optimisation techniques. However, the main advantage of GA techniques is their ability to converge to a global optimum in the case of difficult and multimodal criterion functions. It is important to point out that if the genetic algorithm is used to tune a controller structure we need not know much about the process except very rough bounds of the region of search in the parameter space. These bounds could be arbitrarily loose. Of course, the larger the region of search, the more experiments must be carried out in order to yield the optimal solution.

5.5   Laboratory set-up

The entire experimental set-up is presented in Figure 5.4. The laboratory plant is connected to a PC-486 computer via the Burr-Brown PCI 20000 process interface module with A/D and D/A channels. As a supplement, it is possible also to attach a Honeywell LP 41105-5031 pressure transmitter to the top of the tube in order to sense pressure in the air cushion. Measurement and process control routines are carried out by means of the SIMCOS simulation language which allows for simulation in real-time [34]. SIMCOS can simulate continuous, discrete and hybrid processes and can 'talk' to interface cards and thus send and receive data directly from the process. SIMCOS is based on CSSL syntax and runs under DOS. All the control algorithms discussed above are implemented in their continuous form. Proper 'continuous' behaviour is guaranteed by low sampling times (T_0=0.1 s).

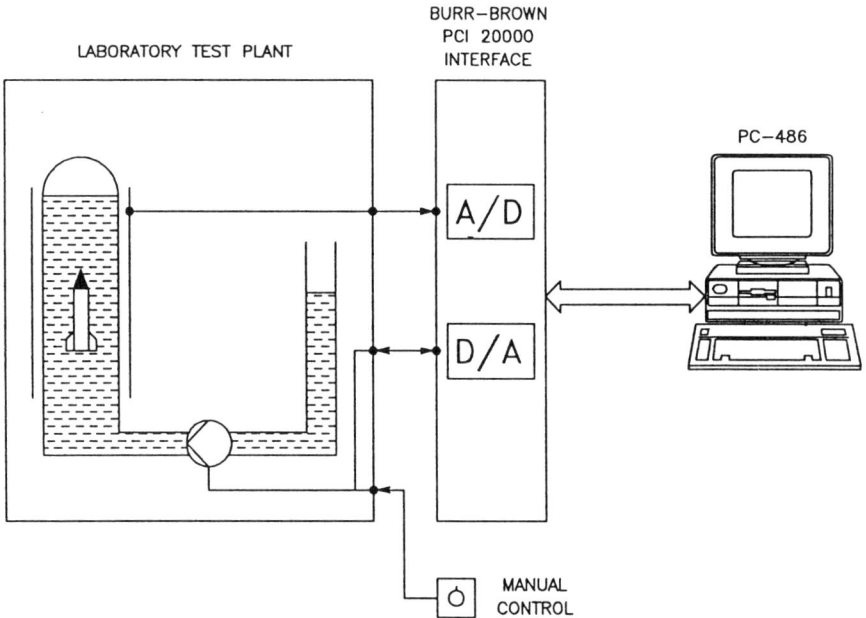

Figure 5.4   The experimental set-up

5.6 Suggested experiments

A closed loop step response is taken as a basis for comparison of performance of the controllers discussed above. For this purpose the system is first stabilised at a working point, after which a step change in the reference signal is applied. In order to check robustness it is important to observe the step responses, not only at the original working conditions (i.e. those for which the linearised model is obtained) but also at degraded working conditions. In our case by decreasing the initial pressure at the top of the tube (i.e. pressure at $u=0$) process gain is increased.

The following experiments are performed:

(1) step response of the LQR with additional integral state. The rocket position and pressure in the air cushion are directly measured. The speed of the rocket is obtained indirectly by filtering the position signal. Since the band of the derivative action of the filter is wider than the closed loop bandwidth, we practically have real speed at our disposal

(2) step response of the LQG controller (LQR with Kalman filter)

(3) step response of the model-based PID controller

(4) an experimental run of an interactive genetic algorithm which serves to tune the PID structure on the assumption that the plant model is not known.

The design of the different controller structures in (1), (2) and (3) is carried out in MATLAB with the aid of the CONTROL Toolbox. The design criteria are for the settling time to be as short as possible, for low overshoot (possibly less than 20%) and good gain and phase margins.

5.7 Illustrative results

5.7.1 Linear quadratic regulator

The original process model (Equation 5.6) is extended with additional integral state, i.e.

$$x = \left[z\,\dot{z}\,p\,\xi\,\right]' , \quad \xi = \int_0^\infty \{r(t) - y(t)\}\,dt \qquad (5.16)$$

where $r(t)$ denotes the reference signal. The weighting matrices Q and R in Equation 5.9 are determined by trials. The choice $Q = \mathrm{diag}(1, 0, 0, 1)$ and $R = [1]$ resulted in controller gain $K = [9.17, 11.3, 1.9 \times 10^{-4}, -1.]$ and the closed loop poles $\{-2, -0.8, -0.45 \pm j0.25\}$ which are considered suitable. The return ratio (see

Figure 5.5) has 60.15° phase margin and infinite gain margin which is a guarantee of good robustness with respect to changes in the process. This is most evident in the fact that step responses at normal and degraded operating conditions are nearly the same (see Figure 5.5).

5.7.2   Linear quadratic regulator with observer (LQG design)

The same state controller is equipped with the observer (only the rocket position is measurable). The Kalman filter approach is used so that the result of weight adjustments is observer gain $L = [10.83, 53.7, 2.28 \times 10^6]$ which yields the observer poles $\{-6.43, -3.2 \pm j4.8\}$. Here the classical rule which recommends that we take the observer poles appropriately faster than the regulator poles has appeared useful from the point of view of robustness. The result is a phase margin of 35° and gain margin of 10.5 dB. Good margins result in minimally degraded step responses in the case of changed operating conditions (Figure 5.6). Note that performance is very similar to that of the full-state controller. The break frequency is $w_B = 3\ \mathrm{s}^{-1}$ which is slightly higher than in the previous case ($w_B = 2.5\ \mathrm{s}^{-1}$).

5.7.3   Model-based PID

The PID is structurally 'poor' compared with the LQG. For example, the PID is not able to introduce as much phase advance as the LQG (compare Figures 5.6 and 5.7) although there are similarities in the corresponding magnitude and phase plots. Obviously, we have less design freedom so that problems stressed in the first section become evident. A relatively good choice ($K_p = 10$, $T_i = 5$, $T_d = 2$) results in 13° of phase margin and 5.7 dB of gain margin, which is worse than for the previous case ($w_B = 3\ \mathrm{s}^{-1}$). The implication on the closed-loop step response is obvious (Figure 5.7). Note also that it was not possible to yield the requirements stated above, i.e. overshoots are too high (>40%).

   Note also that in the analysis we use both Bode and Nyquist plots. The former offer a somewhat clearer picture due to use of a logarithmic frequency scale. However, in the case of unstable systems it is not possible to judge the closed-loop stability solely from the Bode plot. Correct insight can only be gained from the Nyquist plot.

5.7.4   The optimised PID using a GA approach

The problem addressed in proposed experiment (4) is to find the optimal PID parameters which yield as fast as possible a settling time with an overshoot no greater than 10%. The settling time was defined as a moment t_s such that:

$$\left| \frac{y(r) - r}{r} \right| < \varepsilon, \ \forall \, t \geq t_s$$

whereas r denotes the step change in the reference signal.

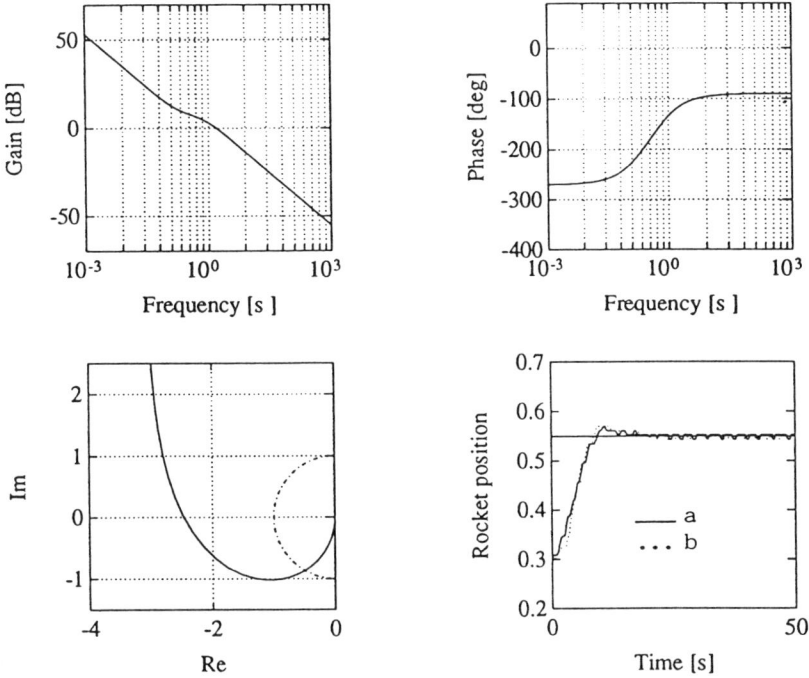

Figure 5.5   *Bode and Nyquist plots for the full-state linear quadratic regulator (LQR)*

Step responses at (*a*) nominal and (*b*) degraded working conditions are depicted

Two series of tuning experiments were performed directly on the test plant. In the first case $\varepsilon = 0.1$, whilst in the second case $\varepsilon = 0.05$. The search space was defined *ad hoc* as a cube $K_p \in [0, 25.4]$, $K_i = K_p/T_i \in [0, 6.3]$, $K_d = K_p * T_d \in [0, 25.4]$. Candidate solutions were encoded as 21-bit strings consisting of three 7-bit sections representing three PID parameter values. They were evaluated by the so-called fitness function [25]. The parameters of the GA were as follows: the number of generations was set to 30, the number of iterations in each generation (population size) was 10, crossover probability was 0.7, the number of crossing sites was 3 and the mutation rate 0.05. The best of the results generated are given in Figure 5.8. Two observations are needed here:

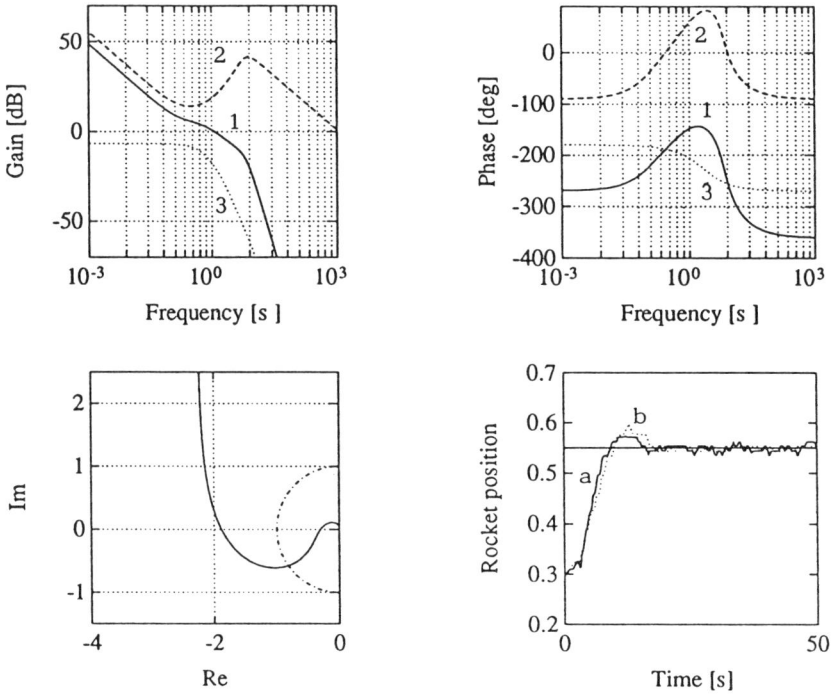

Figure 5.6   Design of the LQG controller

1, 2 and 3 denote Bode plots of the return ratio, LQG transfer function and open-loop plant transfer function, respectively

Step responses at (*a*) nominal and (*b*) degraded working conditions are depicted

First, note that in spite of tough requirements the algorithm has succeeded in finding a PID setting that yields lower settling times and lower overshoots than in the case of model-based controllers (particularly PID). In the previous case we designed controllers on the model whilst GAs design the controller on the process. Since the model is imperfect, some aspects of reality are overlooked by it, but not by the interactive GA!

Secondly, note that the choice of ε substantially influences the GA search. In the case of a loose steady-state boundary (high ε), GAs insist on higher K_p and K_d (fast action!) hence resulting in low overshoots well in accordance with the notes made in the second section. However, in the case of $\varepsilon = 0.05$, the GA could not find feasible solutions up to the 15th generation. In that case the resulting K_p and K_d are smaller than in the case of $\varepsilon = 0.1$, which results in a calm response in the steady state.

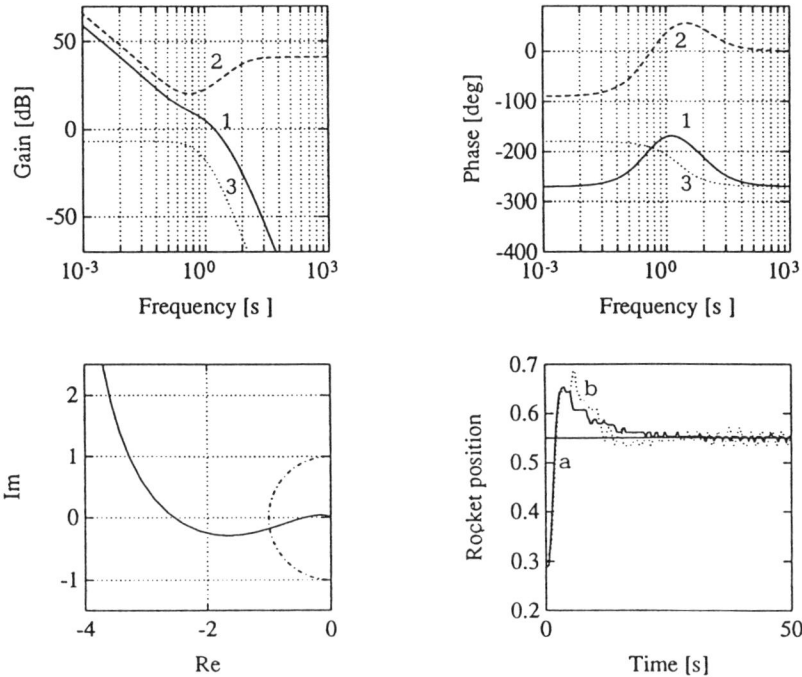

Figure 5.7   *Model-based design of a PID controller*
1, 2 and 3 denote Bode plots of the return ratio, PID transfer function and open-loop plant transfer function respectively
Step responses at (a) nominal and (b) degraded working conditions are depicted

5.8   Conclusions

Several controller structures and design techniques for control of unstable systems have been reviewed and practically demonstrated on a test plant. It has been shown that a full-order LQR yields the best performance and robustness properties. In the case of single measurable output LQG and PID controllers were applied. The LQG is a higher-order controller that gives more opportunities for shaping the loop transfer than PID so that better gain and phase margins can be achieved. However, for a less experienced practitioner the PID might appear more attractive since it is relatively free of the huge theoretical background that comes with LQG, and it has only three tuning knobs. The ultimate aim in model-based design is a good model which might take a great deal of effort. As a remedy, there is direct tuning by means of effective genetic algorithms, which does not require knowledge of the process model. This could yield substantial savings in design time, although the price is a lack of insight into the properties of the control loop

and an excessive number of tuning experiments that must be performed directly
on the process.

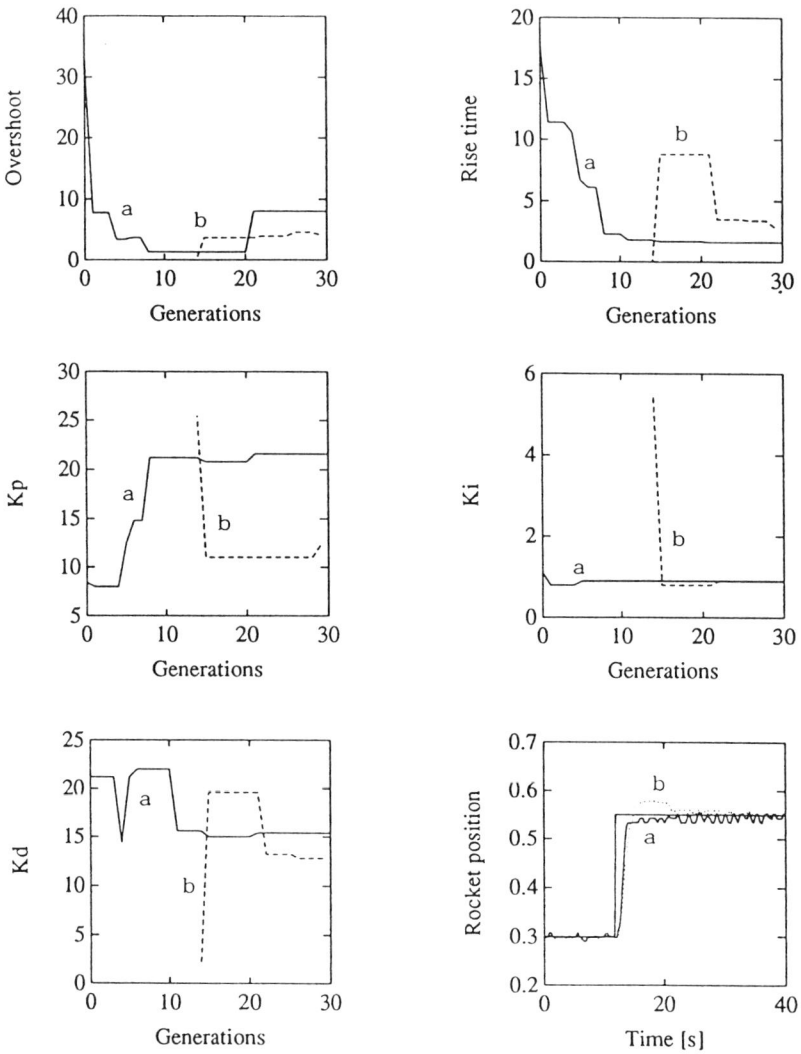

Figure 5.8   Best of the generation results of PID tuning with an interactive
genetic algorithm at different steady state boundaries
(a) ε = 0.1 and (b) ε = 0.05
Step responses of the optimal solutions achieved are also depicted

5.9 Acknowledgment

The author gratefully acknowledges the support of the Slovene Ministry of Science and Technology.

5.10 References

1   SAFONOV, M. G., LAUB, A. J. and HARTMANN, G.: 'Feedback properties of multivariable systems: the role and use of return difference matrix', *IEEE Trans. Automatic Control*, 1981, **26**, pp. 47–65

2   SAFONOV, M. G. and CHIANG, R. Y.: 'CACSD using the state-space L^∞ theory — a design example', *IEEE Trans. Automatic Control*, 1988, **33**, pp. 477–479

3   SHINSKY, F. G.: 'Process control systems: application, design, and adjustment' (McGraw-Hill, New York, 1988)

4   JIANG, J. and DORAISWAMI, R.: 'Convergence analysis of least-squares identification algorithm for unstable systems', *IEE Proc. D*, 1987, **5**, pp. 301–308

5   ISERMANN, R.: 'Identifikation dynamischer Systeme, Band II', (Springer Verlag, München, 1988)

6   LJUNG, L.: 'System identification: theory for the user', (Prentice-Hall, New York, 1987)

7   MA, C. C. H.: 'Unstabilizability of linear unstable systems with input limits', *Proc. ACC*, 1991, **1**, pp. 130–131

8   FREUDENBERG, J. S. and LOOZE, D. P.: 'Right half plane poles and zeros and design trade-offs in feedback systems', *IEEE Trans. Automatic Control*, 1985, **30**, pp. 555–565

9   MACIEJOWSKI, J. M.: 'Multivariable feedback design', (Addison-Wesley, New York, 1989)

10   MIDDLETON, R. H.: 'Trade-offs in linear control system design', *Automatica*, 1991, **27**, (2), pp. 281–292

11   OLSSON, G. and PIANI, G.: 'Computer systems for automatic control', (Prentice-Hall, New York, 1992)

12   ISERMANN, R.: 'Digital control systems', (Springer Verlag, New York, 1981)

13   ASTRÖM, K. J. and WITTENMARK, B.: 'Computer controlled systems', (Prentice-Hall, New York, 1984)

14   FRIEDLAND, B.: 'Control system design: an introduction to state-space method', (McGraw-Hill, New York, 1986)

15   KWAKERNAAK, H. and SIVAN, R.: 'Linear optimal control systems', (Wiley, New York, 1972)

16   FRANKLIN, G. F. and POWELL, J. D.: 'Digital control of dynamic systems', (Addison-Wesley, Reading MA, 1980)

17   ATHANS, M. and FALB, P. L.: 'Optimal control: an introduction to the theory and its applications', (McGraw-Hill, New York, 1966)

18   PREUβ, H. -P.: 'Prozeβmodellfreier PID-Regler-Entwurf nach dem Betragsoptimum', *Automatisierungstechnik*, 1991, **39**, pp. 15–22

19   KRAUS, T. W. and MYRON, T. J.: 'Self-tuning PID controller uses pattern recognition approach', *Control Engineering*, June 1984, pp. 108–111

20   ASTRÖM, K. J., HANG, C. C., PERSSON, P. and HO, W. K.: 'Towards intelligent PID control', *Automatica*, 1992, **28**, (1), pp. 1–9

21   D'AZZO, J. J. and HOUPIS, C. H.: 'Feedback control system analysis and synthesis', (McGraw-Hill, New York, 1966)

22   GOLDBERG, D .E.: 'Genetic algorithms in search, optimization and machine learning', (Addison-Wesley, Reading MA, 1989)

23   DAVIS, L.: 'Handbook of genetic algorithms', (Van Nostrand Reinhold, New York, 1991)

24   RAWLINS, G. J. E.: 'Foundations of genetic algorithms', (Morgan Kaufmann, San Mateo, 1991)

25   FILIPIÈ, B. and JURICIC, Ð.: 'An interactive genetic algorithm for controller parameter optimization', *Prepr. Int. Conf. Neural Networks and Genetic Algorithms*, Innsbruck, 1993, pp. 458–462

26   DEJONG, K.: 'Adaptive system design: a genetic approach', *IEEE Trans. Systems, Man. and Cybernetics*, 1980, **10**, (9), pp. 566–574

27   PORTER, B., MOHAMED, S. S. and JONES, A. H.: 'Genetic tuning of digital PID controllers for linear multivariable plants', *Proc. 2nd European Control Conference*, 1993, Groningen, Vol. 3, pp. 1392–1396

28   WANG, P. and KWOK, D. P.: 'Optimal design of PID process controllers based on genetic algorithms', *Prepr. 12th IFAC World Congress*, 1993, Sidney, Vol. V, pp. 261–264

29   JOHNSON, C. D.: 'The "unreachable poles" defect in LQR theory: analysis and remedy', *Int. J. of Control*, 1988, **47**, (3), pp. 697–709

30   DOYLE, J. C.: 'Guaranteed margins for LQG regulators', *IEEE Trans. Automatic Control*, 1978, **23**, pp. 756–757

31   DOYLE, J. C., and STEIN, G.: 'Robustness with observers', *IEEE Trans. Automatic Control*, 1979, **24**, pp. 607–611

32   SCHMITENDORF, W. E.: 'Design of observer-based robust stabilizing controllers', *Automatica*, 1988, **24**, pp. 693–696

33   BARMISH, B. R. and GALIMIDI, A. R.: 'Robustness of Luenberger observers: linear system stabilized via nonlinear control', *Automatica*, 1986, **22**, (4), pp. 413–423

34   ZUPANCIC, B.: 'SIMCOS language for simulation of continuous and discrete dynamic systems', University of Ljubljana, 1991

Chapter 6

Control of temperature and heat flow rate: the problem of delays

P. Zítek

6.1 Introduction

This chapter is devoted to a significant problem in feedback control, namely the problem of delays in the loop. The longer the delays in signal processing, the poorer the control action is likely to be, including loss of the ability to control at all, i.e. the loss of stability. Many phenomena cause different types of delays, of which the most frequent are:

— transport, i.e. a change occurring at one point in the process is detected elsewhere.
— distributed process parameters, i.e. properties of continuous spatial variables.
— latent changes in the process.

In the following, the essential features of the control problems resulting from delays in the control loop are explained and some basic approaches for solutions are outlined. A method of complex plane representation of the control problem is presented and its application is demonstrated on a heat transfer process represented by a specific laboratory set-up. Three options of control, namely the classical PID control loop, a feed-forward application and state feedback, are presented, compared and discussed.

6.2 Control problem statement

Any kind of mass transport or energy transfer over a distance brings about the phenomenon of delay in the associated control system, i.e. the consequent effect appears some time after its cause. Essentially, the longer the distances between system components, the larger the resulting delays will be. Many methods of control system analysis and design do not specifically take account of these delays, despite the frequent occurrence of delay phenomena in process dynamics. However, as any feedback controller derives its actuating signal from the output response of the process, any kind of input-output delay will influence the control action in a strong negative sense. Feedback control loops with considerable delays usually have unsatisfactory properties and sometimes these delays are so long that acceptable control action cannot be achieved by single loop control alone. Thermal processes in general, and heat exchange processes in particular, are typical physical processes with long delays. For this reason a heat transfer control laboratory set-up is used as an example here.

Delays in a process may be represented in various forms. A very simple one is dead-time delay, or transport lag, described by the formula:

$$r(t) = x(t - \tau_D) \tag{6.1}$$

where x, r are the original and delayed variables, respectively, and τ_D is a constant. A more general delay relationship may be expressed by the Stieltjes integral:

$$r(t) = \int_0^T x(t - \tau) dh(\tau) \tag{6.2}$$

where t is a variable, T is its maximum and $h(t)$ describes the delay distribution. The step responses of both of these relationships are shown in Figure 6.1; the delays in Equations 6.1 and 6.2 may be considered as *concentrated* and *distributed* delays, respectively. The longer the parameters t_D or T, the more time is lost in controller reaction. From this point of view, solving the problem of control loop delay becomes a problem of minimising the negative influence of this lost time on the control of the process.

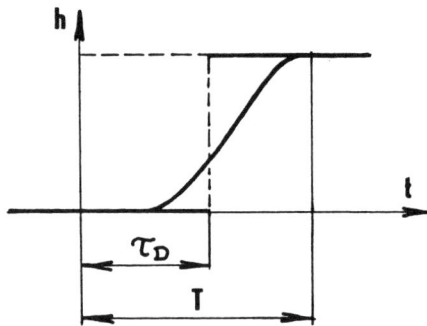

Figure 6.1   Step responses of systems with delay

6.2.1 Basic approaches to controlling processes with delay

If the delay in the process is too long to achieve an acceptable control action in a single feedback loop (block R in Figure 6.2) there are three basic approaches to manage the problem:

— to apply a feed-forward compensation of the dominant input(s);
— to use state variables to achieve a less delayed actuating signal (i.e. state feedback);
— to attempt to predict continuation of the process in the controller action (e.g. Smith predictor controller).

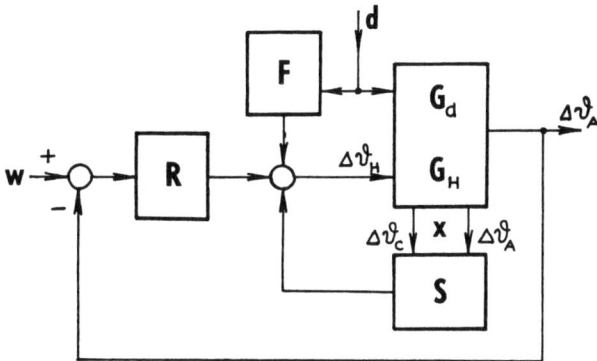

Figure 6.2   Control schemes to cope with the time delay

The feed-forward compensation is represented by the block F and the state feedback by the block S in Figure 6.2. The state feedback approach provides a simple solution where state variables are measurable. In contrast, state variables which are not easily measured require, as shown in the previous chapter, the so-called state observer in the control loop [1]. The Smith predictor arrangement is used only rarely; for more information see, for example, Reference 2. For feed-forward compensation it should be stressed that only the disturbance or load which is measured, and whose influence can be identified, can be compensated. Any other disturbing process input can not be counteracted.

6.2.2 Control problem example

The control problem of heating a building may be used as a typical example of a thermal control process in which delays are important. The purpose is to keep the interior temperature at the required constant value in spite of the substantial and irregular influence of weather conditions outside. This very slow thermal process may be modelled in an accelerated form by a laboratory set-up as sketched in Figure 6.3, where the heated room is represented by a vessel A (of volume V_A).

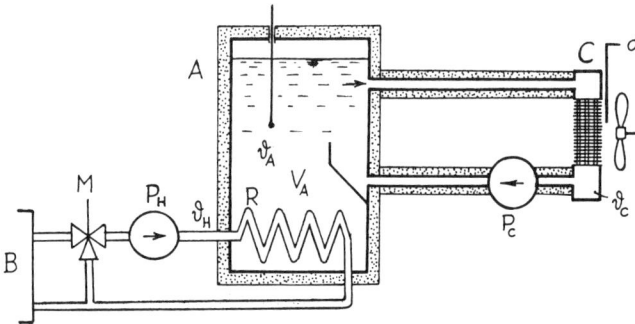

Figure 6.3   Process of heating a building (laboratory set-up)

Room heating is represented by a hot water radiator H and the outside weather influence by water circulating through an external cooler C, driven by a pump. Hot water for heating is prepared in an electric boiler and the heating performance is controlled by adjusting a mixing valve. The essential control problems are as follows:

— Due to a low heat transfer rate (i.e. insulation) the interior temperature responds to any sudden heat exchange imbalance with a considerable delay;
— In addition, the interior water temperature varies randomly within a certain range due to gravitational circulation;
—Changing the boiler temperature is not suitable due to its poor dynamics and also to its hastening influence on the corrosion process.

6.3   Process model analysis

To design a sound operational control system a model must be found to analyse the dynamic properties of the process. In fact, the heat exchangers mentioned represent a process with distributed parameters. However, after merging the heat capacities of the vessel A and cooler C, respectively, the following heat balance relationships may be formulated for them according to Figure 6.1:

$$M_A c \frac{d\vartheta_A}{dt} = K_H\left[\vartheta_H(t-\tau_H) - \vartheta_A(t)\right] - Q_C c\left[\vartheta_A(t) - \vartheta_C(t-\tau_C)\right] \qquad (6.3)$$

$$M_C c \frac{d\vartheta_C}{dt} = Q_C c\left[\vartheta_A(t-\tau_A) - \vartheta_C(t)\right] - dK_C\left[\vartheta_C(t) - \vartheta_0(t)\right] \qquad (6.4)$$

The symbols used denote:

ϑ_A	average water temperature inside vessel A
ϑ_H	effective temperature of heating water
ϑ_C	average water temperature inside cooler C
M_A	mass of the contents of vessel A
M_C	mass of the contents of cooler C
c	specific thermal capacity
Q_C	rate of flow of the cooling circulation
K_H	effective heat exchange rate of the radiator
K_C	effective heat exchange rate of the cooler (screen fully open)
d	relative cooler screen opening
τ_H	dead time of the heater response
τ_A, τ_C	circulation transport times between the vessel and the cooler

Obviously, the equations obtained are neither linear nor ordinary. Amongst others, one of their non-linearities results from the fact that the heat exchange parameters K_H, K_C are not constant but depend on their respective rates of flow. The next specific feature of Equations 6.3 and 6.4 is the time shift where τ_H, τ_A, τ_C express the respective delays. After linearisation and application of a Laplace transform, Equations 6.3 and 6.4 may be transformed into the form

$$T_A s\Delta\vartheta_A(s) = \kappa_H(\Delta\vartheta_H(s)e^{-s\tau_H} - \Delta\vartheta_A(s)) - \Delta\vartheta_A(s) + \Delta\vartheta_C(s)e^{-s\tau_C} \qquad (6.5)$$

$$T_C s\Delta\vartheta_C(s) = \Delta\vartheta_A(s)e^{-s\tau_A} - \Delta\vartheta_C(s) - \kappa_C d_0\Delta\vartheta_C(s) - \kappa_C\Theta\Delta d \qquad (6.6)$$

where the operator Δ denotes the respective deviations from the reference values, and the coefficients are

$$T_A = \frac{M_A}{Q_C}, \quad T_C = \frac{M_C}{Q_C}, \quad \kappa_H = \frac{K_H}{Q_C c}, \quad \kappa_C = \frac{K_C}{Q_C c}, \quad \Theta = \vartheta_{CO} - \vartheta_O$$

In the laboratory set-up the heat exchangers were designed with minimal volumes. Their parameters have the following values:

$$M_A c = 418 \text{ kJ K}^{-1}, \ M_C c = 0.42 \text{ kJ K}^{-1}, \ Q_C c = 0.209 \text{ kJ s}^{-1} \text{ K}^{-1}$$
$$K_H = 0.0253 \text{ kJ s}^{-1} \text{ K}^{-1}, \ K_C = 0.0029 \text{ kJ s}^{-1} \text{ K}^{-1}, \tau_H = 6\text{s}, \tau_A = \tau_C = 4\text{s}$$

The flow rates through the heat exchangers and K_H and K_C are considered as constants in Equations 6.5 and 6.6.

Based on Equations 6.5 and 6.6, a linear process model results in the block diagram of Figure 6.2. The dynamics are represented by transfer functions $G_H(s)$

and $G_d(s)$ for the heating water temperature and cooling deviations, respectively. Unlike the usual rational fraction form these transfer functions are transcendental:

$$G_H(s) = \frac{\kappa_H \left(T_C s + 1 + \kappa_c d_0\right) e^{-s\tau_H}}{\left(T_A s + 1 + \kappa_H\right)\left(T_C s + 1 + \kappa_c d_0\right) - e^{-s(\tau_A + \tau_C)}} \qquad (6.7)$$

$$G_d(s) = \frac{-\kappa_c \Theta e^{-s\tau_C}}{\left(T_A s + 1 + \kappa_H\right)\left(T_C s + 1 + \kappa_c d_0\right) - e^{-s(\tau_A + \tau_C)}} \qquad (6.8)$$

This form of transfer functions means the controlled process does not belong to the class of minimal-phase systems. Hence, although they are only of second order, their phase angle is not limited as frequency tends to infinity due to the exponentials $e^{-s\tau_H}, e^{-s\tau_C}$ in their numerators. A rather complicated problem arises in the case of any application using the poles and zeros of $G(s)$ because their number is unlimited [3].

6.4   Control system design

6.4.1   Technical approach and discussion

The model in Equations 6.3 and 6.4 provides a necessary basis for discussing the control problems. First, which manipulating variable is the most suitable for controlling heating performance? Adjusting the flow rate through radiator R is very easy but its value influences the heat exchange rate K_H in an intricate and non-linear way. The unsuitability of any way of changing the outlet temperature of boiler B has already been mentioned. The most convenient solution remaining is to control the heating performance by adjusting the inlet water temperature ϑ_H into radiator H by mixing the flow from boiler B with the return flow. This is carried out by a mixing valve as shown in Figure 6.3.

The second question is that of deciding by which controlled variable the mixing valve should be adjusted. Since the original task is to maintain the interior temperature ϑ_A, one standard solution is to arrange a normal feedback loop adjusting the valve according to ϑ_A deviation.

However, as mentioned above, this is not the only method that can be applied. First, since the change of cooling effect happens to be the dominant disturbance of the process, there is a good opportunity here to employ the measured screen position d to control ϑ_H. However, the change in cooling effect may result not only from the displacement d but also from changes in air temperature, humidity or other parameters. For this reason the state feedback principle is a more perfect solution, particularly if the so-called anisochronic state concept [4] is applied.

Using this approach the mixing valve position is adjusted according to both temperatures ϑ_A and ϑ_C (as state variables) with suitable controller parameters.

6.4.2 Single feedback loop with a PID controller

Consider a PID controller with a transfer function

$$R(s) = \frac{r_0}{1 + T_M s}\left[\frac{1}{T_i s} + 1 + s T_d\right] \tag{6.9}$$

in the feedback loop outlined in Figure 6.2. The symbols for controller parameters denote:

r_0 proportional gain
T_i integral time
T_d derivative time
T_M temperature sensor time constant

The corresponding control system dynamics are expressed by the equation

$$\{1 + G_H(s)R(s)\}\Delta\vartheta_A(s) = G_d(s)\Delta d \tag{6.10}$$

Using the model in Equations 6.7, 6.8 and 6.9 with the parameters given above, the following system equation results:

$$\left[(20s + 1.121)(2s + 1.014) - e^{-8s} + 0.121(2s + 1.014)\frac{e^{-6s}}{1 + T_M s}\left(\frac{A_1}{s} + A_2 + A_3 s\right)\right]$$

$$\times \Delta\vartheta_A(s) = -\kappa_C \Theta e^{-4s}\Delta d(s)$$

$$\tag{6.11}$$

where $A_1 = r_0 / T_i$, $A_2 = r_0$, $A_3 = r_0 T_d$.

6.4.3 State feedback control arrangement

In Equations 6.3 and 6.4 the temperatures ϑ_A, ϑ_C may be considered as state variables. In fact, they should not be considered as state variables as commonly introduced in standard control theory, e.g. Reference 1, because of the delay relations in the two equations. To introduce ϑ_A, ϑ_C as state variables it is necessary to apply so-called anisochronic (functional) state space [4,5] to treat the delays in Equations 6.3 and 6.4. Briefly, this means to consider these equations as state equations and, as for the system state, to consider a trajectory segment instead of a point only. Nevertheless, the state feedback control principle remains the same: the S controller works as a combined feedback proportional to the deviations of both state variables, as follows:

$$\Delta\vartheta_H = -\begin{bmatrix} K_1, K_2 \end{bmatrix}\begin{bmatrix} \Delta\vartheta_A \\ \Delta\vartheta_C \end{bmatrix} \tag{6.12}$$

where K_1, K_2 are the feedback gains.

After inserting Equation 6.12 into the transform Equations 6.5 and 6.6, the control system model acquires the form

$$T_A s\Delta\vartheta_A(s) = K_H\left(-K_1\Delta\vartheta_A(s)e^{-s\tau_H} - K_2\Delta\vartheta_C(s)e^{-s\tau_H} - \Delta\vartheta_A(s)\right)$$
$$-\Delta\vartheta_A(s) + \Delta\vartheta_C(s)e^{-s\tau_C} \tag{6.13}$$

$$T_C s\Delta\vartheta_C(s) = \Delta\vartheta_A(s)e^{-s\tau_A} - \Delta\vartheta_C(s) - \kappa_C d_0\Delta\vartheta_C(s) - \kappa_C\Theta\Delta d \tag{6.14}$$

It is not difficult to understand why we need to involve the temperature ϑ_C in the controller action in this case: it responds much more quickly to the input disturbance (a change in cooling) than the interior temperature ϑ_A. Positioning the controller sensors as near as possible to the source of the imbalance is the basic way to reduce the harmful influence of delay.

6.5   Controller parameter assignment

Control loops with delays are particularly sensitive to controller parameter setting. Some propensity to lose stability is often observed in their behaviour, for which reason the design of controller action has to be carried out more carefully than usual. However, as soon as any form of delay in the process dynamics has occurred, there is little option but to apply control theory. One difficulty we are faced with here is a diminishing choice of analytical methods resulting from the transcendental character of both the transfer functions and the Laplace transform equations describing any system with delay; see Equations 6.5–6.8, for instance.

A Laplace transform model of the thermal control system discussed may be generally expressed in the form

$$M(s)\Delta\vartheta_A(s) = N(s)\Delta d(s) \tag{6.15}$$

where, obviously, $M(s)$, $N(s)$ are no longer polynomials but transcendental functions. This also means that, the usual algebraic analytical methods are no longer valid in this class of problems. The key point of this complication is that, unlike an algebraic equation, the transcendental one

$$M(s) = 0 \tag{6.16}$$

is satisfied by an infinite number of solutions (roots *M*-zeros) in the complex domain. None of these zeros, of course, may be placed in the right half plane.

On the other hand, the function $M(s)$ is transcendental in *s* only, not with respect to the controller parameters. Therefore if a fixed complex number $s=p_i$ is inserted into Equation 6.16, a linear algebraic equation results

$$M(p_i, K_1, K_2, ...) = 0 \tag{6.17}$$

for unknown controller parameters K_1, K_2.

6.5.1   *PID controller setting assignment*

A Laplace transform equation for the control system with a PID controller was obtained as Equation 6.11. For the process parameters given above, the system's characteristic function is:

$$M(s) = 80s^4 + (85.04 + 0.24 A_3 e^{-6s})s^3 + (24.79 - e^{-8s} + [0.24 A_2 + 0.122 A_3]e^{-6s})s^2$$
$$+ (1.134 - e^{-8s} + [0.24 A_1 + 0.122 A_2]e^{-6s})s + 0.122 A_1 e^{-6s}$$

$$\tag{6.18}$$

Two basic tasks may be solved using this function. First, a trial or assigned parameter setting may be tested by $M(s)$, after inserting $s=jw$, $M(jw)$. Secondly, some required zeros $M(s)$ may be prescribed and corresponding controller parameters assigned. However, this second task is not as simple as it looks. With an infinite number of *M* zeros, no finite number of 'stable' and 'well-damped' *M* zeros (prescribed and achieved) can guarantee either stability or good properties for the control system. For this reason it is necessary to repeatedly combine both the above mentioned procedures in the following way:

— to compute the controller parameters for prescribed p_i
— to check the resulting dynamics by means of $M(jw)$

The full procedure is as follows:

(1)   For an initial trial setting of control coefficients, e.g. A_1=2.5, A_2=20, A_3=40, to compute the characteristic function $M(s)$ for $s=jw$; see Figure 6.4, contour A.
(2)   Evaluating the contour $M(jw)$ to check the system's stability and dynamic properties. Since the complete argument increment of the contour A in Figure 6.4 is

$$\lim_{\omega \to \infty} \Delta \arg M(jw) = 4\pi / 2 = 360° \qquad (6.19)$$

the system is stable, however this stability is not very strong. The natural oscillations at $w=0.108\,\text{s}^{-1}$ are only weakly damped, because their damping ratio is $\delta = 0.126$ (the distance contour $M(jw)$ – s-origin measured in the w scale $wd=0.0136$).

(3)   With respect to the $M(jw)$ contour, to estimate the attainable frequency w_R of the desired control process and its damping rate d. For example, in the case discussed, a required frequency of $w_R=0.04\,\text{s}^{-1}$ and a good damping rate $d=0.4$ may be tried.

(4)   After inserting M zeros corresponding to the required w_R and d, a set of equations for controller parameters can be obtained and then solved.

(5)   The new controller parameters obtained from the solution of (4) do not yet necessarily assure an acceptable control process. Hence a new $M(jw)$ contour needs to be computed. If it satisfies the stability conditions, a simulation experiment is used for final validation of the result.

(6)   If the result obtained is not acceptable (e.g. negative controller parameters, unstable control of process, too large control error etc.), a new frequency ω_R should be selected and procedures (3)–(5) repeated.

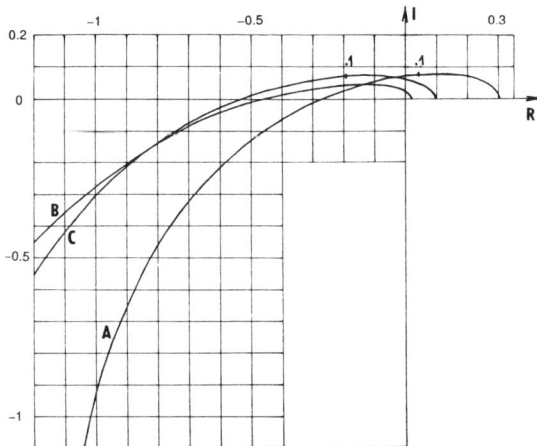

Figure 6.4   Plots of M(jw). PID control

To achieve relatively strong damping with $d = 0.4$ at the required frequency $w_R = 0.04\,\text{s}^{-1}$ three controller parameters need to be adjusted. However, the prescribed w_R and d represent only two conditions. The third can be obtained in the form of a Lagrange multiplier relation. Inserting the required

$$p_{1,2} = \omega_R\left(-\delta \pm j\right) \tag{6.20}$$

into $M(s)$, two equations for the real and imaginary parts, $M = R + jI$, are obtained with A_1, A_2, A_3 to be solved

$$R\left(p_{1,2}\right) = 0.1288496 A_1 - 0.0012355 A_2 - 0.00019961 A_3 - 0.0117954 = 0$$
$$I\left(p_{1,2}\right) = -0.0206525 A_1 + 0.0060462 A_2 - 0.00013717 - 0.0369 = 0 \tag{6.21}$$

If the new $A_{1,2,3}$ are considered as readjustments $DA_{1,2,3}$ of the original $A_{10} = 2.5$, $A_{20} = 20$, $A_{30} = 40$ the Equations 6.21 acquire the form

$$1.288496\Delta A_1 - 0.012355\Delta A_2 - 0.0019961\Delta A_3 = -2.776342$$
$$-0.206525\Delta A_1 + 0.06462\Delta A_2 - 0.0013717\Delta A_3 = -0.269056 \tag{6.22}$$

Instead of the third required M zero it is possible to apply the well-tried ratio integration/derivative time constant from the Ziegler–Nichols rules

$$\frac{T_i}{T_d} = \frac{A_2^2}{A_1 A_3} = m \tag{6.23}$$

where m is a constant usually equal to 4. From this rule the following Lagrange multiplier condition may be derived:

$$\Delta A_1 = \frac{\partial A_1}{\partial A_2}\Delta A_2 + \frac{\partial A_1}{\partial A_3}\Delta A_3 = \frac{A_2}{2A_3}\Delta A_2 - \left(\frac{A_2}{2A_3}\right)^2 \Delta A_3 \tag{6.24}$$

where A_2/A_3 is taken from the initial trial setting ($A_2/A_3 = 20/40 = 0.5$).
From the Equation set 6.21–6.24 a solution is obtained:

$$\Delta A_1 = -2.297, \quad \Delta A_2 = -12.605, \quad \Delta A_3 = -13.672$$

which means the parameters $A_1 = 0.203$, $A_2 = 7.395$, $A_3 = 26.328$, satisfying both the stability condition (contour B in Figure 6.4) and the damping requirement as well (response B in Figure 6.5). Unfortunately, the resulting control process is rather slow, prompting the question of whether such strong damping is necessary. For rather lower damping, e.g. $d = 0.3$ and a higher frequency $w_R = 0.05~\text{s}^{-1}$, the procedure introduced above yields higher values of parameters $A_1 = 0.803$, $A_2 = 12.113$, $A_3 = 37.221$, with stable $M(jw)$, contour C in Figure 6.4, and a quicker response C in Figure 6.5. Obviously, slower responses commonly lead to higher values of control error.

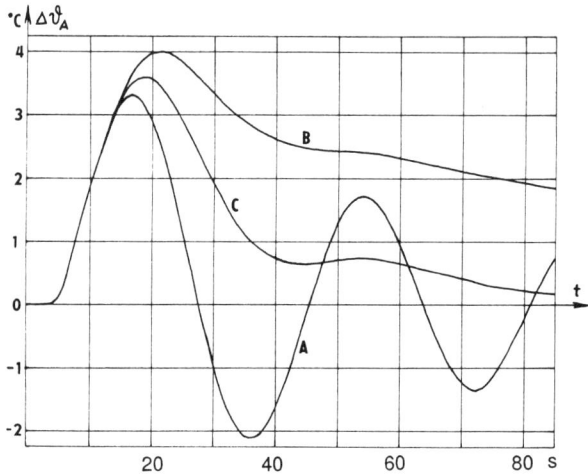

Figure 6.5   Step responses. PID control

6.5.2   State feedback parameter assignment

The parameters of the state feedback controller are to be assigned in a similar way as for PID. The characteristic function corresponding to Equations 6.13 and 6.14 is as follows:

$$M(s) = s^2 + s\left(0.006053K_1e^{-6s} + 0.560577\right) + 0.0030539K_1 +$$
$$+ 0.02828 + 0.003012K_2e^{-10s} - 0.02488e^{-8s} \qquad (6.25)$$

As a first attempt, dynamic trial parameters $K_1 = K_2 = 2.5$ have been selected. The corresponding $M(jw)$ contour is shown in Figure 6.6 as curve A. From this experiment the complex conjugate pair may be chosen: $p_{1,2} = -0.1 \pm j0.2$ as the prescribed M zeros (damping ratio $d = 0.5$). Inserting these p_i into Equation 6.25, the following equations are obtained for the real and imaginary parts, respectively:

$$0.367264K_1 - 0.3040719K_2 = 5.616088$$
$$0.335908K_1 + 0.744484K_2 = 12.74632 \qquad (6.26)$$

with the solution $K_1 = 21.949$, $K_2 = 7.206$. Also this result does not guarantee an acceptable control process since it does not specify the infinite set of possible further M zeros. Only the corresponding contour B in Figure 6.6 proves not only the stability but also the absence of oscillations with weaker damping than required.

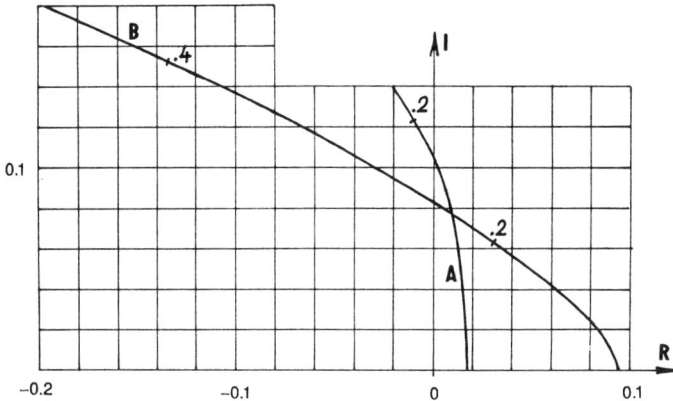

Figure 6.6   Plots of M(jw). State feedback options

6.6   Suggested experiments

Thermal processes are often too slow for demonstration of their control problems in real-time in an educational setting. The aim of the laboratory set-up described is to present the control phenomena in a more accelerated manner and to offer a clear experimental verification of various control solutions. The original heating control process is one of those processes which is strongly influenced by delays and is not easily investigated by standard models based on ordinary differential equations or transfer functions of the rational fraction type.

The set-up makes it possible to investigate the influence of changing the delays and other dynamic parameters. It is effected using various lengths of connecting tubes between the vessel A and cooler C and, particularly, by changing the flow rate of the circulation. In this way we can see the growing difficulties in accomplishing acceptable control caused by increasing delays in the loop. On the other hand it demonstrates how an important beneficial effect is gained by applying an auxiliary control signal in addition to the main control error, which is less burdened and distorted by delays.

As far as controller implementation is concerned, it should be emphasised that the latest compact controllers (in the set-up Siemens SIPART DR 20 was used) can realise not only the PID function but also more complex functions with auxiliary signals, such as state feedback or cascade control.

6.7   Conclusions

The dynamic behaviour of control systems with delays or after-effects is essentially more complex in character than can be observed in systems without any delay, which are considered in standard control theory. In the case of transcendental transfer functions the usual algebraic methods in control theory (based on ordinary differential equations) fail when applied in this problem area. The characteristic equation for any control system with delays has to allow an infinite set of solutions, which is why it cannot be substituted by an algebraic one. However, in spite of this property, the control process does not look too complicated, as it is governed almost completely by only so-called predominant M zeros. Other M zeros lying in the higher frequency range are not visible in the system response.

Computer simulation is a good means for final verification of controller design and adjustment. Nevertheless, the complex plane provides a more effective tool for finding an efficient controller structure and its settings to satisfy the specification required. The method presented of a characteristic function $M(s)$ works with only the left-hand operator of the control system equation; however, in fact, the slow dynamics of a system with consequent after-effects does not allow prompt tracking of transients, which systems without delays are able to manage through feedback control. From this point of view, the main control task is to restrict the natural tendency to oscillate and to search for any possible opportunity to apply an auxiliary signal which is closer to the causes of the imbalance to be overcome by control. It is also necessary to recognise an important difference between systems with and without delays. While the continuation of a delay-free system's motion is determined only by its instantaneous state variable values, the motion of a system with delays always results from its recent history, the effect of which is given by the length of its delays. Owing to this property these systems are also referred to as hereditary ones. Their different causality of motion represents the main reason to apply special methods of control system design.

6.8   References

1   ACKERMANN, J.: 'Sampled-data control systems. Analysis and synthesis, robust system design', (Springer-Verlag, Berlin, 1985)

2   RAY, W. H.: 'Advanced process control', (McGraw-Hill, New York, 1981)

3   MYSHKIS, A. D.: 'Linear differential equations with delayed argument', (Nauka, Moscow, 1972, in Russian)

4   ZÍTEK, P: 'Anisochronic modelling and stability criterion of hereditary systems', *Problems of control and information theory*, 1986, **15**, (6), pp. 413–423

5   ZÍTEK, P.: 'Control synthesis of systems with hereditary properties', *3rd IEEE Conference on Control Applications*, Glasgow, 1994

6.9   Further reading

MALEK-ZAVAREI, M. and JAMSHIDI, M.: 'Time-delay systems: analysis, optimization and applications', (North-Holland, 1987)

HALE, J.: 'Theory of functional differential equations', (Springer, New York, 1977)

ZÍTEK, P.: 'Anisochronic generalization of dynamic system state', *in IFAC 4th symposium on automatization in mining, mineral and metal proc.*, Helsinki, 1983

Chapter 7

Inverted pendulum control

P.M. Frank and N. Kiupel

7.1 Introduction

Using state observers it is possible to reconstruct the state vector of observable systems using a mathematical model which is excited by the input and output signals.

First, linear system description is introduced in this experiment. Design and calculation of the Luenberger identity observer, using the pole placement method, is then explained. To put the observer to a practical test, the algorithm was implemented on a microcomputer system and applied, for purposes of state observation, to the 'inverted pendulum' laboratory model.

By assignment of different observer poles, the dynamic behaviour of the observer can be varied. The resulting effects of these variations on the observer estimates are measured. By comparison of the state estimates with the actual states in the physical system, the respective estimation errors can be determined and evaluated. The results demonstrate which prerequisites for a meaningful state observation must be fulfilled for the 'inverted pendulum' example.

7.2 Theoretical foundations

Time-invariant physical systems can generally be described by nth order nonlinear differential equations. If the differential equation is linearised about a pre-assigned operating point and the differentials are defined as the new state vector, a system of first order linear differential equations describing the state equation

$$\dot{x}(t) = Ax(t) + Bu(t) \qquad (7.1)$$

is obtained, the output equation being expressed by

$$\dot{y}(t) = Cx(t) + Du(t) \tag{7.2}$$

According to Reference 1 the solution of the state equation is

$$x(t) = \Phi(t - t_0)x(t_0) + \int_{t_0}^{t} \Phi(t - \tau)Bu(\tau)d\tau \tag{7.3}$$

where the matrix

$$\Phi(t) = e^{At} = I + At + \frac{A^2 t^2}{2!} + \dots$$

is called the fundamental or transition matrix. The states of the sampled data systems can be computed by integration over the sampling period T. Assuming a piecewise constant input signal

$$u(t) = u(kT) \qquad kT \le t \le (k+1)T \tag{7.4}$$

the discrete time sampled data model of the system is given by

$$x((k+1)T) = A_D(T)x(kT) + B_D(T)u(kT) \tag{7.5}$$

where

$$A_D(T) = \Phi(T); \qquad B_D(T) = \int_0^T A_D(\tau)Bd\tau \tag{7.6}$$

These matrices should be computed only once, because they are independent of the input signal. For the sampled data system, the output equation (7.2) remains unchanged.

If a state space system description is given, the system can be controlled by feedback of the state vector to the system input. If the system is controllable (see the definition in Reference 2 or 3), the poles of the open loop system are shifted to stable locations by a suitably selected feedback matrix F. A prefilter, which is described by the matrix V, ensures in the case $p=m$ that in the steady state the output vector y is identical to the setpoint vector w.

The input signal u generated by the control system is now computed by the setpoint signal w and the state quantity Fx which is fed back

$$u(kT) = Vw(kT) - Fx(kT) \tag{7.7}$$

This relation yields the system description of the closed loop control system

$$x((k+1)T) = (A_D - B_D F)x(kT) + B_D Vw(kT) \tag{7.8}$$

$$y(kT) = (C - DF)x(kT) + DVw(kT) \tag{7.9}$$

7.3 The 'inverted pendulum' system

The 'inverted pendulum' system consists of a cart which can be moved along a metal guiding bar. An aluminium rod with a cylindrical weight is fixed to the cart by an axis. The cart is connected by a transmission belt to a drive wheel. The wheel is driven by a current-controlled d.c. motor which delivers a torque proportional to the acting control voltage u_S such that the cart is accelerated.

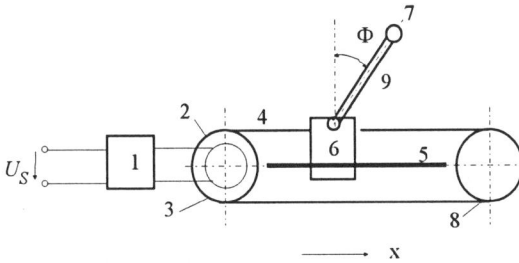

Figure 7.1   *Principle scheme of the 'inverted pendulum' model*
1   Servo amplifier
2   Motor
3   Drive wheel
4   Transmission belt
5   Metal guided bar
6   Cart

The following quantities are measured in the pendulum system:

1. The position of the cart, by means of a circular-coil potentiometer which is fixed to the driving shaft of the motor;
2. The velocity of the cart, by means of a tacho generator which is also fixed to the motor;
3. The angle of the pendulum rod, by means of a layer potentiometer which is fixed to the pivot of the pendulum.

The inverted pendulum is commercially available (see, for instance, Reference 5).

7.3.1   *Mathematical model of the inverted pendulum*

In the following, a mathematical model for the inverted pendulum system is to be derived. For the following explanations we refer to Reference 5. In Figure 7.2 the overall system is divided into the two systems 'cart' and 'pendulum'. The acting forces are also shown. If the mass of the pendulum is M_1 and r denotes the position of the cart, the following force acts horizontally on the bottom point of the pendulum:

$$H = M_1 \frac{d^2}{dt^2}(r + l_s \sin\Phi) \qquad (7.10)$$

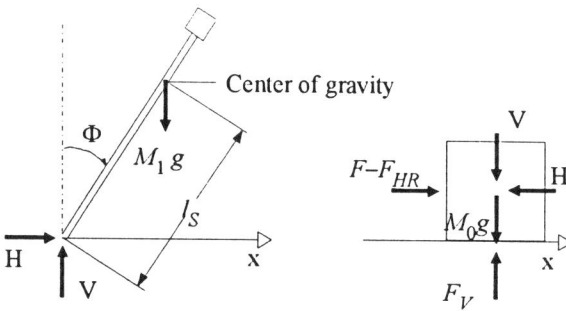

Figure 7.2   Free body diagram of the pendulum and the cart

This force is due to the acceleration of the centre of gravity. The vertical component of the force can be computed as

$$V = M_1 \frac{d^2}{dt^2}(l_s \cos\Phi) + M_1 g \qquad (7.11)$$

The angular momentum conservation law yields, for the rotary motion of the rod about the centre of gravity,

$$\Theta_s \frac{d^2\Phi}{dt^2} = Vl_s \sin\Phi - Hl_s \cos\Phi - C\frac{d\Phi}{dt} \qquad (7.12)$$

where Θ_s denotes the inertia moment of the pendulum rod with respect to the centre of gravity. C is the friction constant of the pendulum. For the cart system, the equation of motion can be written as

$$M_0 \frac{d^2 r}{dt^2} = F - H - F_r \frac{dr}{dt} \qquad (7.13)$$

where the mass of the cart is denoted by M_0. The velocity-proportional friction constant is called F_r. The force acting via the transmission belt is represented by F. Differentiation of the trigonometric functions yields

$$H = M_1 (\ddot{r} + l_s \ddot{\Phi} \cos \Phi - l_s \dot{\Phi}^2 \sin \Phi) \qquad (7.14)$$

and

$$V = -M_1 l_s (\ddot{\Phi} \sin \Phi + \dot{\Phi}^2 \cos \Phi) + M_1 g \qquad (7.15)$$

After some straightforward manipulations, one obtains the nonlinear differential equations

$$\Theta \ddot{\Phi} + C \dot{\Phi} - M_1 l_s g \sin \Phi + M_1 l_s \ddot{r} \cos \Phi = 0 \qquad (7.16)$$

and

$$M \ddot{r} + F_r \dot{r} + M_1 l_s (\ddot{\Phi} \cos \Phi - \dot{\Phi}^2 \sin \Phi) = F \qquad (7.17)$$

These equations describe the mathematical model of the pendulum in the form of a system of coupled differential equations. The following abbreviations have been used

$$\Theta = \Theta_S + M_1 l_s^2; \quad M = M_0 + M_1 \qquad (7.18)$$

These equations are the basis for the derivation of a mathematical control model of the inverted pendulum.

7.3.2 Description of the linearised system in the state space

To complete the description of the 'pendulum' system, a linearisation about a suitable operating point must be performed. The first step is the introduction of a state vector

$$x = \begin{bmatrix} x_1 & x_2 & x_3 & x_4 \end{bmatrix}^T = \begin{bmatrix} r & \Phi & \dot{r} & \dot{\Phi} \end{bmatrix}^T \qquad (7.19)$$

and an input signal $u = F$. Transforming into the form

$$\dot{x} = f(x, u)$$

yields:

$$\dot{x}_1 = f_1(x,u) = x_3 \tag{7.20}$$

$$\dot{x}_2 = f_2(x,u) = x_4 \tag{7.21}$$

$$\dot{x}_3 = f_3(x,u) = \beta(x_2)(a_{32}\sin x_2 \cos x_2 + a_{33}x_3 + a_{34}\cos x_2 x_4 + a_{35}\sin x_2 x_4^2 + b_3 u) \tag{7.22}$$

$$\dot{x}_4 = f_4(x,u) = \beta(x_2)(a_{42}\sin x_2 + a_{43}\cos x_2 x_3 + a_{44}x_4 + a_{45}\cos x_2 \sin x_2 x_4^2 + b_4 \cos x_2 u) \tag{7.23}$$

where the abbreviations

$$\beta(x_2) = \left(1 + \frac{N^2}{N_{01}^2}\sin^2 x_2\right)^{-1} \tag{7.24}$$

$$N = M_1 l_S \tag{7.25}$$

$$N_{01}^2 = \Theta M - N^2 \tag{7.26}$$

have been used, and the coefficients a_{ij}, b_i can be easily derived.

In the second step, the operating point, about which the linearisation is to be performed, is defined. It is given by the condition

$$x_A = \begin{bmatrix} k_1 & 0 & 0 & 0 \end{bmatrix}^T \tag{7.27}$$

The set of equations (7.20–7.23) are developed into a Taylor series which is terminated after the first element.

$$\Delta\dot{x} = \left.\frac{\partial f}{\partial x}\right|_{\substack{x=x_A \\ u=0}} \cdot \Delta x + \left.\frac{\partial f}{\partial u}\right|_{\substack{x=x_A \\ u=0}} \cdot \Delta u \tag{7.28}$$

The computation of the differential quotient yields for the linearised system matrix

$$\frac{\partial f}{\partial x}\bigg|_{\substack{x=x_A \\ u=0}} = \begin{bmatrix} 0 & 0 & 1 & 0 \\ 0 & 0 & 0 & 1 \\ 0 & a_{32} & a_{33} & a_{34} \\ 0 & a_{42} & a_{43} & a_{44} \end{bmatrix} \equiv A_A; \quad \frac{\partial f}{\partial u}\bigg|_{\substack{x=x_A \\ u=0}} = \begin{bmatrix} 0 \\ 0 \\ b_3 \\ b_4 \end{bmatrix} \equiv b_A \qquad (7.29)$$

The deviations from the operating point are introduced as the new state variables and input signals $\Delta x = x$ and $\Delta u = u$, respectively. The state equation can then be written as

$$\dot{x} = A_A x + b_A u \qquad (7.30)$$

This equation constitutes the linear state space description of the inverted pendulum. Thus, the prerequisites for the application of the controller and observer design techniques introduced in Section 7.2 are satisfied.

7.3.3   Normalisation of the state equations

The 'inverted pendulum' system is equipped with three sensors, by which the position of the cart, the velocity of the cart and the angle of the pendulum rod are measured. The fourth state variable, i.e. the angular velocity of the pendulum rod, is estimated by a state observer such that state feedback can be performed.

Since the sensors deliver a voltage which is proportional to the mechanical quantities to be measured, it is necessary to normalise the system with respect to the electrical quantities. The normalisation can be understood as a regular transformation

$$x_n = Nx \qquad (7.31)$$

and

$$u = K_F u_S \qquad (7.32)$$

of the non-normalised state vector x and the non-normalised input signal u where

$$N = \text{diag}[n_{ii}] \quad i = 1,...4 \qquad (7.33)$$

is a diagonal matrix. The term K_F denotes the factor relating the input voltage of the servo amplifier to the force acting on the cart. Another substitution results in

$$N^{-1}\dot{x}_n = A_A N^{-1} x_n + b_A K_F u_S \qquad (7.34)$$

Solving this equation for $\dot{x}_n$ yields the normalised system description

$$\dot{x}_n = NA_A N^{-1} x_n + Nb_A K_F u_S \qquad (7.34)$$

with the normalised system matrix

$$A_n = NA_A N^{-1} \qquad (7.35)$$

and the normalised input matrix

$$b_n = Nb_A K_F \qquad (7.36)$$

The elements n_{11} to n_{33} are identical to the factors relating the electrical to the mechanical quantities. Since the angular velocity is not measured directly, it is possible to assign the element n_{44} arbitrarily, on the condition that the numerical value of the voltage for the angular velocity does not exceed the maximum voltage of the D/A-converter. The element has been selected to be

$$n_{44} = n_{22} \qquad \text{sec} \qquad (7.37)$$

7.3.4   Control and disturbance signal observation in the 'inverted pendulum'

It has already been pointed out that only three of altogether four state variables are measured at the system 'pendulum'. In the following, the design of a reduced order observer for the estimation of the fourth state variable, as well as of the disturbance signal u_{S0} is introduced. The reduced order observer is distinguished from the identity observer in the sense that it does not estimate output signals which are already being measured. Thus the reduced order observer has only the order $n–p$.

Since in this case the disturbance signal u_{S0} is to be estimated additionally, a model augmentation is required which describes the properties of the disturbance. In the mathematical model the quantity u_{S0} has been assumed to be constant. Thus it can be interpreted as a solution of the differential equation

$$\dot{x}_5 = 0 \qquad (7.38)$$

This signal acts via the factor b_3 on the quantity $\dot{x}_3$ and via the factor b_4 on the quantity $\dot{x}_4$, such that the augmented system can be written in the form

$$\begin{bmatrix} \dot{x} \\ \dot{x}_s \end{bmatrix} = \begin{bmatrix} A_A & b_A \\ 0^T & 0 \end{bmatrix} \begin{bmatrix} x \\ x_s \end{bmatrix} + \begin{bmatrix} b_A \\ 0 \end{bmatrix} u \tag{7.39}$$

An observer employing all three measured signals is designed for the augmented system. Figure 7.3 displays the overall system consisting of state feedback control and disturbance compensation.

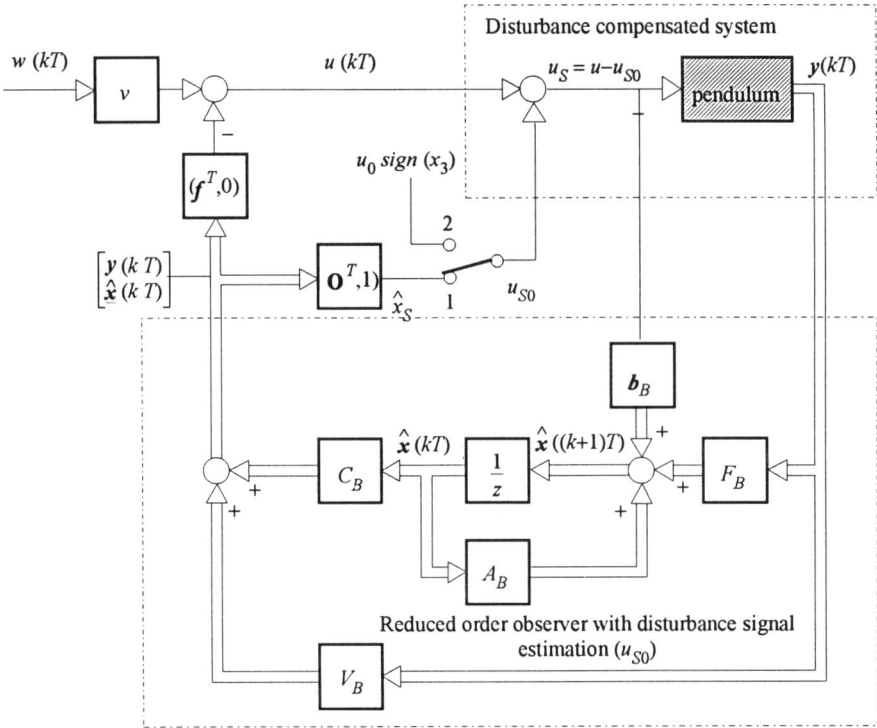

Figure 7.3   *Block diagram of the controlled pendulum with state feedback and reduced order observer*

In switch position 1, the nonlinear friction effects are compensated by the disturbance signal observer. In switch position 2, they are compensated in an analogue fashion. According to Figure 7.3, the equation for the reduced order observer is given by

$$\hat{x}((k+1)T) = A_B\hat{x}(kT) + F_B y(kT) + b_B u_S(kT) \tag{7.40}$$

$$\begin{bmatrix} y(kT) \\ \hat{x}(kT) \end{bmatrix} = C_B \hat{x}(kT) + V_B y(kT) \tag{7.41}$$

For the purpose of computing the unknown matrices, the system is discretised and separated into two subsystems. The 5×5 system matrix can be divided into the four submatrices A_{11}, A_{12}, A_{21}, A_{22}. The vectors b_1 and b_2 are built by the input vector.

$$\begin{bmatrix} y((k+1)T) \\ \hat{x}_B((k+1)T) \end{bmatrix} = \begin{bmatrix} A_{11} & A_{12} \\ A_{21} & A_{22} \end{bmatrix} \begin{bmatrix} y(kT) \\ \hat{x}_B(kT) \end{bmatrix} + \begin{bmatrix} b_1 \\ b_2 \end{bmatrix} u_S(kT) \tag{7.42}$$

$$\hat{x}_B(kT) = \begin{bmatrix} \hat{x}_4(kT) \\ \hat{x}_5(kT) \end{bmatrix}$$

The measurable state variables (x_1, x_2, x_3) of this system are identical to the output signals and are described by the vector y.

If a reduced order observer is designed for this system

$$\hat{x}_B((k+1)T) = \left(A_{22} - LA_{12} \right)\hat{x}_B(kT) + L\left(y\left((k+1)T\right) - A_{11}y(kT) - b_1 u_S(kT) \right)$$
$$+ A_{21}y(kT) + b_2 u_S(kT) \tag{7.43}$$

via the change of variables

$$z(kT) = \hat{x}_B(kT) - Ly(kT) \tag{7.44}$$

the observer equation

$$z((k+1)T) = A_B z(kT) + F_B y(kT) + b_B u_S(kT) \tag{7.45}$$

where
$$A_B = A_{22} - LA_{12}; \quad F_B = A_B L + A_{21} - LA_{11}; \quad b_B = b_2 - Lb_1$$

is obtained. The matrix L has been selected such that the poles of the observer are located at $s_{1,2} = -100 \ 1/s$ or $z_{1,2} = -0.0498$, respectively. The remaining matrices eventually result in

$$V_B = \begin{bmatrix} I \\ L \end{bmatrix}$$

Mathematical techniques for the design of reduced order observers are described in References 3 and 1. Before the feedback matrix is finally computed, the poles of the open loop system will be calculated. They can be determined by

$$\det(sI - A_A) = 0$$

From the values of the poles, it can be seen that one has an unstable system which can be stabilised by means of the state feedback shown in Figure 7.3.

7.4   Suggested experiments

This system is very suitable for learning control techniques. Therefore one can calculate equations for the system, or a part of it. From the equations it is possible to draw the block diagram of the system, either of the 'cart' system or the 'inverted pendulum' system. The next steps are the design of the Luenberger observer and, finally, the controller, as a state space feedback controller. As an experiment the friction constant should be identified from the step response of the system.

7.5 Illustrative results

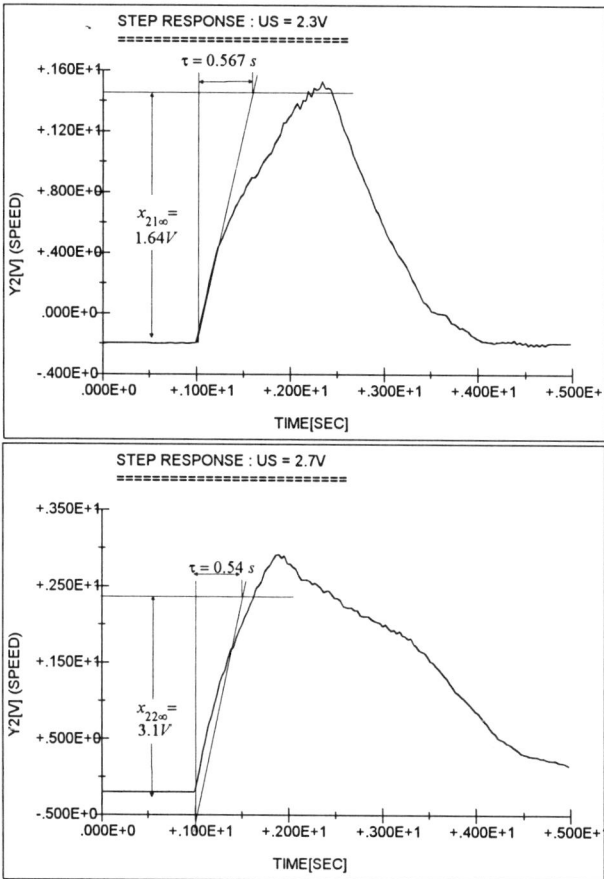

Figure 7.4   *Step response of the 'cart' system*
           upper: $u_S = 2.3$ V
           lower: $u_S = 2.7$ V

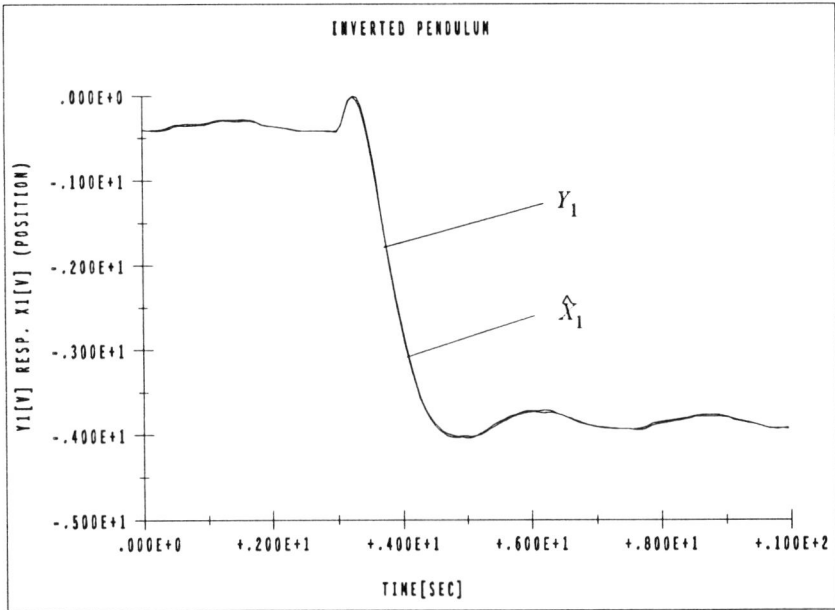

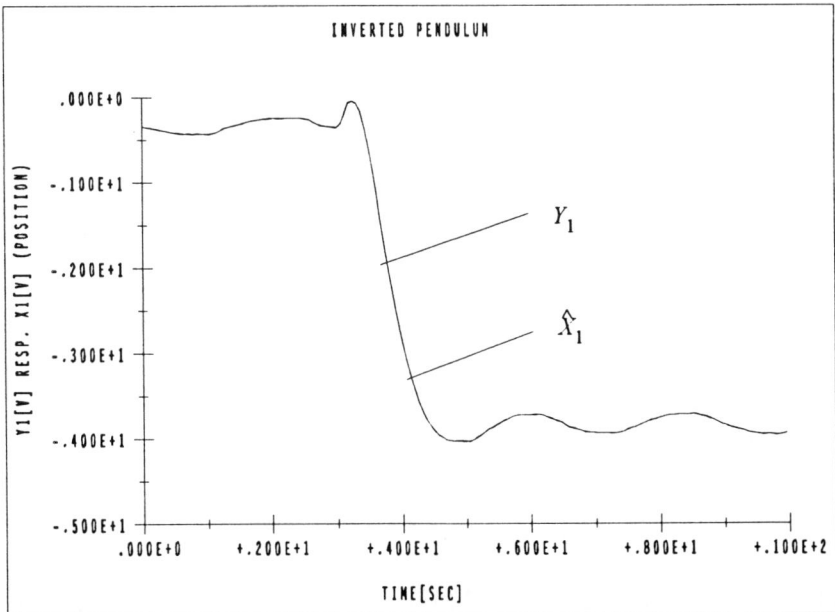

Figure 7.5   *Observation of the position for the 'inverted pendulum' system*
upper: Observer poles at: $s = -6 \, 1/s$
lower: Observer poles at: $s = -20 \, 1/s$

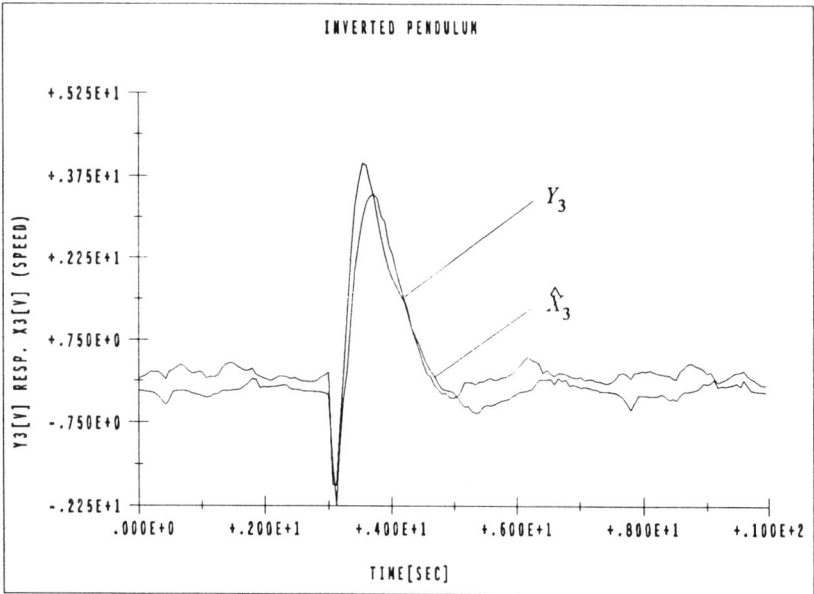

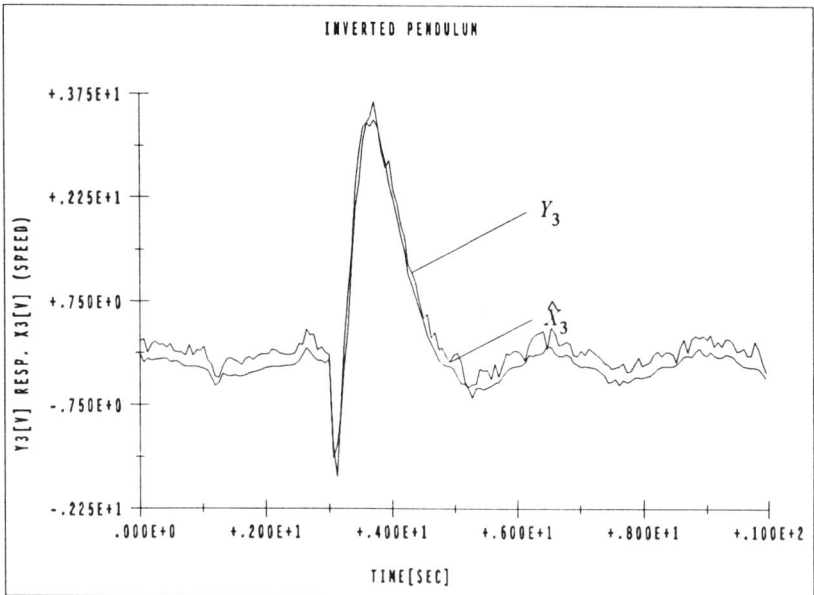

Figure 7.6   *Observation of the speed for the 'inverted pendulum' system*
upper: Observer poles at: $s = -6\ 1/s$
lower: Observer poles at: $s = -20\ 1/s$

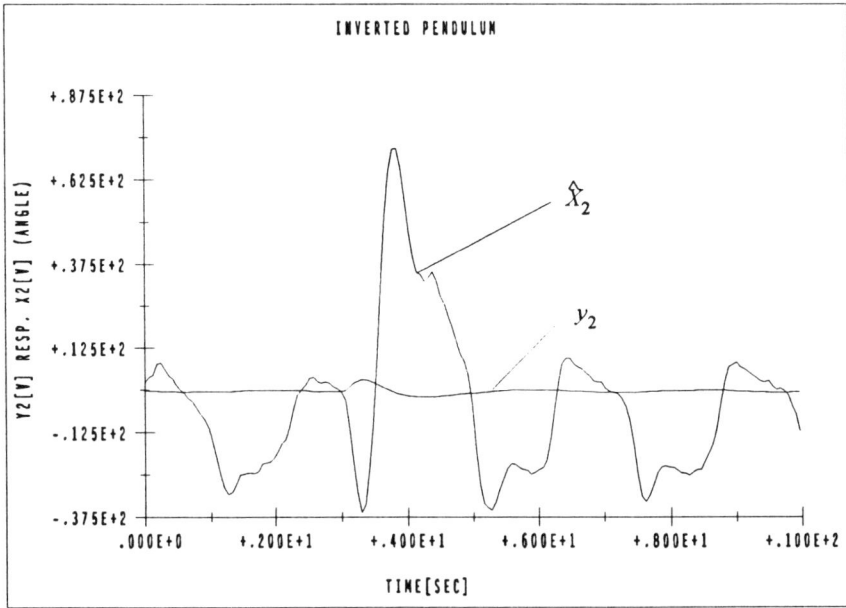

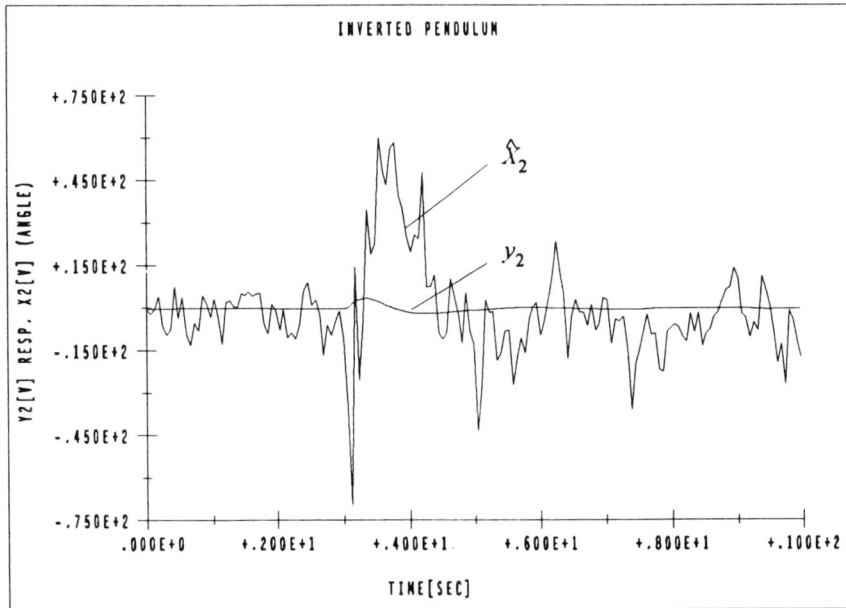

Figure 7.7   *Observation of the angle for the 'inverted pendulum' system*
upper: Observer poles at: $s = -6\,1/s$
lower: Observer poles at: $s = -20\,1/s$

7.6   Conclusions

In this chapter we have described the 'inverted pendulum' system as an example for teaching control techniques, including a short introduction to the theoretical background. The system equations have been derived to provide a mathematical model of the inverted pendulum. With these tools a closed loop system has been calculated using state space techniques to keep the inverted pendulum upright. Another task is to identify the friction constant of the system.

The different theoretical methods applied to the system mean that it provides a good application example within a control course aimed at teaching modern control methods.

7.7   References

1   FOLLINGER, O.: Regelungstechnik, 2nd Ed., (Elitera-Verlag, Berlin, 1978)

2   UNBEHAUEN, H: Regelungstechnik II, 4th Ed., (Vieweg - Verlag, Braunschweig, 1987)

3   ACKERMANN, J: Abtastregelung, (Springer - Verlag, Berlin, 1972)

4   MORI, S., NISHIHARA, H. and FURUTA, K.: Control of unstable mechanical system, Control of pendulum, *Int. J. Control*, 1976, **23**, No. 5, pp. 673–692

5   Amira GmbH: Manual for the laboratory set-up of the inverted pendulum control, Bismarckstr. 67, D-47057 Duisburg, Germany, 1993

Chapter 8

Disturbance rejection

P. Albertos and J. Salt

8.1 Introduction

Traditionally, there are two different ways to show the need for a control system: to track a reference signal at either a higher power level or a distant location, or to keep a variable at a set point in the presence of external disturbances or changes in the process. Typical examples of the first case, so-called **servosystems**, are feedback amplifiers or steering systems. On the other hand, **regulation systems** are always present in the process industries.

Although it is clear that both problems can be solved within a common framework, most control system design methodologies rely on controlled system behaviour under changes in the reference or set points.

The purpose of this chapter is to deal with some industrial control problems where disturbance counteraction is the most relevant issue and to review the suitability of classical control solutions to deal with this particular viewpoint.

In a control system, disturbances appear at different points. They may be classified as input, internal or measurement disturbances. Filtering, detection and cancellation are possible approaches to counteract their effect. In the next section, a typical industrial control problem is presented and the various sources of disturbances are analysed. The basic options to deal with these disturbances are then discussed. An easily implemented laboratory set-up will display most of the key issues to solve these control problems.

8.2 Control problem

In most control system design approaches, the specifications are related to the response to changes in the reference, and closed-loop behaviour is analysed without consideration of the effect of external disturbances. However, disturbance rejection specifications are very old [1] and related to feedback control properties.

If the disturbance is measurable, a feedforward controller can be envisaged to counteract the disturbance effect, usually in a steady-state framework.

Let us consider the welding machine depicted in Figure 8.1. To obtain a uniform welding surface, a constant feeding speed of the stick is required, assuming that there is also a constant relative speed between the tool and the surface. The classical control set-up tries to keep the wheel A's angular speed $w(t)$ constant but, owing to wheel wear, the stick feeding speed $v(t)$ is not constant. A periodic disturbance appears. Its frequency is related to the working speed but, in general, no position sensing is implemented in the wheel A.

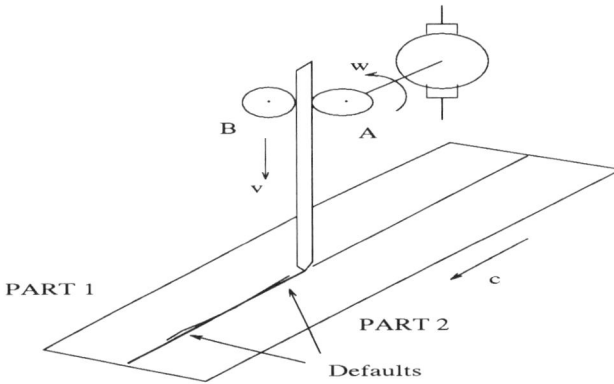

Figure 8.1   Welding process

This is not the only disturbance in the welding process. Working in a hard environment, some measurement noise is unavoidable and the welding load could be variable. A physical block diagram is shown in Figure 8.2.

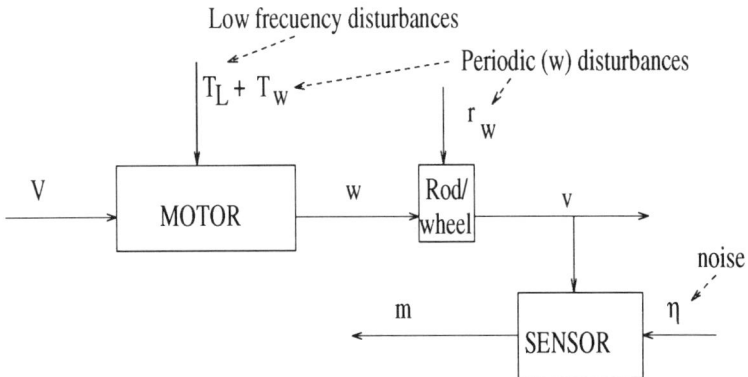

Figure 8.2   Speed measurement

The purpose of this chapter is to analyse basic control system design techniques as they are applied to the specific problem of disturbance rejection. Disturbances will be characterised by their frequency content, in particular their frequency range and magnitude. Other techniques, such as those based on robust controllers or H_∞ specifications, will not be developed in this chapter.

According to these constraints, the proposed methodology is:

(1)   to measure the disturbance and implement a feedforward controller cancelling its effect;

(2)   to detect the disturbance and counteract it by means of a feedback controller; or

(3)   to reduce its relevance either by the controller's integral action or avoiding it feeding back.

The first approach requires disturbance measurement. In the welding control application, some information about the disturbance is available, but its actual value is not. The second approach can be implemented by building a disturbance observer. This requires a good *a priori* knowledge of both the process model and the disturbance model. Alternatively, using a band-pass filter, only knowledge of the disturbance frequency is required. Once the disturbance is estimated one way or another, a control action is generated to counteract it.

The last approach is to consider the disturbance effect on the controlled closed-loop system and to compute an additional control action to reduce it. This approach is simpler but is only useful for disturbance frequencies out of the controlled process range of interest, i.e. for high frequency measurement noise or low frequency load disturbances. In this case the approach is, respectively:

• to filter the measurement noise using a low-pass filter;
• to reduce the disturbance effect by a high-gain, integral action controller.

The application studied covers most of these problems. The following scenarios are considered: high frequency speed measurement noise, low frequency load torque and medium range frequency disturbances.

Both classical and modern control design techniques are used. Although disturbance rejection is the primary goal, the effects of the filter and controller on the input-output reference specifications are also analysed.

A set of practical results obtained at the laboratory using a d.c. motor, controlled by a personal computer with a commercial data acquisition card, is reported. These results allow a comparison between different design procedures, showing the great variety of techniques available to solve industrial problems.

8.3   Technical background

Let us consider a single-input single-output, disturbed linear continuous time-invariant system as depicted in Figure 8.3, where $u(t)$ is the plant input, $s(t)$ is the plant output, $y(t)$ is the measurement signal and $d_1(t)$, $d_2(t)$ and $d_3(t)$ are input, process and measurement disturbances, respectively. $G_i(s)$ are subprocesses transfer functions.

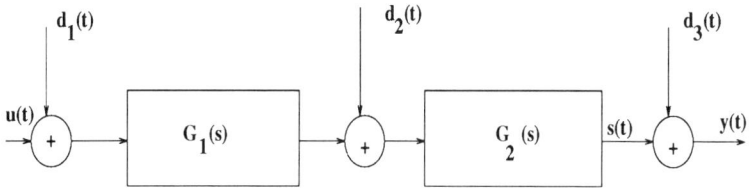

Figure 8.3   Disturbed plant model

For $d_1(t)$ and $d_2(t)$, if they were measurable, a feedforward controller could be designed. If $G_1(s)$ is strictly proper, a control signal

$$u(t) = -\left[d_1(t) + \frac{1}{G(s)} d_2(t) \right]$$

will cancel their effect. If this is not the case, a control

$$u(t) = -\left[d_1(t) + \frac{1}{k_1} d_2(t) \right]$$

where k_1 is the static gain of $G_1(s)$, will avoid the steady-state errors for step disturbances [2]. However, disturbances are not, in general, measurable and they must be counteracted by feedback control actions. In the following, they are assumed to be characterised by their frequency content and generated and modelled by autonomous systems.

The frequency range of these disturbances is quite distinct. The frequency range of $d_3(t)$, the measurement disturbance or noise, is usually higher than that of the desired output signal, $s(t)$, but if it is an input or process disturbance, the range of frequencies is similar to that of the plant.

To take the proper disturbance rejection actions, some information about the disturbances should be available. Two approaches are considered. First, the output/disturbance transfer function, $s(t)/d_i(t)$, of the closed-loop controlled plant is specified and a feedback controller is designed. Second, the disturbance is estimated and then an opposite control action applied.

The first approach provides a structured control solution, although it can distort the desired output/reference transfer function. The second tries to avoid the inability of the disturbance to be measured, but does not avoid the noninvertibility of the transfer functions. Thus, it would mainly be of interest to compensate input disturbances.

8.3.1 Disturbance filtering

The disturbance filtering may be computed directly, inside the control loop, or indirectly, computing the ideal output filter to be mathematically converted into a feedback filter, a complex one. In the first case, two different locations are possible. If the filter is in the feedback path it tries to avoid any disturbance effect on the controlled system. On the other hand, if it is placed in the forward path it tries to generate disturbance-corrective actions.

8.3.1.1 Forward path filtering (low frequency disturbances)

The typical scenario to apply inner loop filtering of the disturbance in the forward path is the case of constant or low frequency disturbances.

If the controller $R(s)$, as shown in Figure 8.4, involves an integral action, the output/disturbance transfer function, $s(t)/d_i(t)$, will present zero static gain, suppressing the possible disturbance error at low frequencies [4]. Some additional constraints may be imposed on the controller in order also to obtain an acceptable transient behaviour. This approach is not suitable for output disturbances. The controlled system's response to reference changes is also affected.

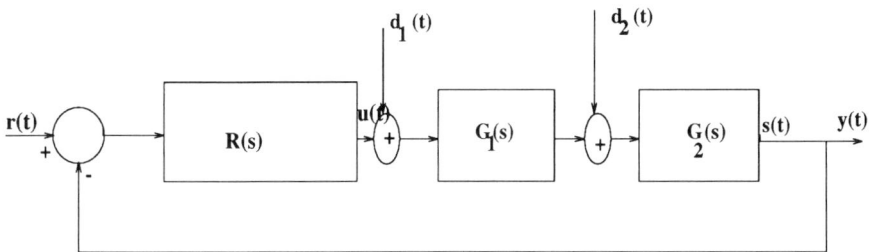

Figure 8.4   Closed-loop controlled plant

8.3.1.2 Feedback path filtering (measurement noise)

Let us now assume a noisy measurement, that is $d_1(t) = d_2(t) = 0$ and $d_3(t) \neq 0$. The measured controlled output $y(t)$ will present this high frequency component. We can compute the appropriate measurement filter $F(s)$ to be implemented in the feedback path, as shown in Figure 8.5, to provide the 'noise-free' feedback $m(t)$ [5].

The measurement filter may be designed as a low-pass filter to suppress the additional high frequency noise. In this way, the plant input will not be disturbed and the plant output, $s(t)$, will be unaffected by this noise. On the other hand, this noise will be evident in the output measurement, $y(t)$.

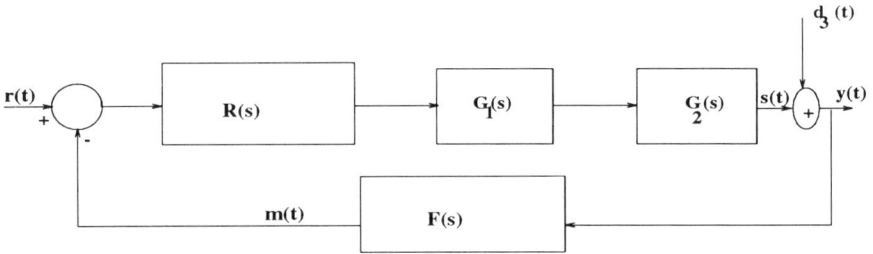

Figure 8.5   Closed-loop control by filtered measurement

If the disturbance is an internal one, such as $d_2(t)$ in Figure 8.3, the previous filter will not give an undisturbed plant output, $s(t) = y(t)$. In this case, external disturbance filtering may be achieved out of the loop. Theoretically, a cascade filter $F(s)$ will delete the disturbance frequency content of the output, giving a better filtered output. But this signal is required not after the filter but at the process output. In order to obtain a clean plant output, the equivalent inner loop $F(s)$ can be derived. Let us assume a controlled system as depicted in Figure 8.3. The equivalent in-loop filter $F(s)$ will be:

$$F(s) = \frac{1 + R(s)G_1(s)G_2(s) - \overline{F}(s)}{R(s)G_1(s)G_2(s)\overline{F}(s)}$$

However, the controller is very complex and, although the frequency components of the disturbance are cancelled, the transient behaviour will also be changed.

8.3.2 Disturbance estimation

If the disturbance is an input disturbance, it may be counteracted. The proposed approach is to estimate this disturbance and to generate an opposite control action.

To estimate the disturbance, the two approaches suggested are by disturbance state estimation or by filtering and forecasting [6]:

• **Disturbance detection.** The main idea is to extract the disturbance-related information from the measured signal. If the disturbance frequency range is delimited and known, a band-pass filter will provide the first component of the disturbance. As a delay will appear, a phase-lead filter will shift the estimated disturbance, at least in the steady state. This requires a precise knowledge of the disturbance frequency.

- **Disturbance state estimation.** With the assumption of simple disturbance models, like pure sinusoidal disturbances, they can be modelled by an autonomous signal generator, as shown in Figure 8.6.

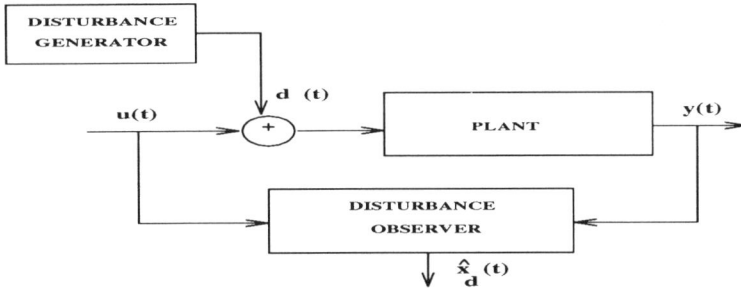

Figure 8.6   Disturbance observer

An augmented system state vector [2] can be defined and a reduced-order observer can be designed to estimate the state of the disturbance generator. The disturbance is then obtained by the appropriate state combination.

$$\hat{d}\big[(k+1)T\big] = C_d v\big[(k+1)T\big]$$

$$v\big[(k+1)T\big] = A_d v(kT)$$

$$v(kT) = K\big[x(kT) - Ax\big[(k-1)T\big] - Bu\big[(k-1)T\big]\big] + (A_d - KBC_d)v\big[(k-1)T\big]$$

where

$\hat{d}$ represents the estimated value of the disturbance;
x and u are the state and the input to the process, respectively;
(A, B) and (A_d, C_d) are the process and disturbance state description matrices;
v is the disturbance state; and
K is the gain matrix of the observer.

In this case, the estimated disturbance will also be delayed and a phase-lead filter will predict the current value of the disturbance. The disturbance frequency is just one parameter of the disturbance generator model. If the frequency slowly varies with time, an adaptive scheme may be implemented, based on the estimation of this parameter.

8.4   Laboratory set-up

The general approaches described above have been applied to design the position control of a laboratory welder stick plant. The stick is fed by a d.c. motor through a carrying wheel which is eccentrically shaped due to wear.

The laboratory set-up consists of:

- a d.c. motor (Quanser Consulting [3]);
- a commercial data acquisition card (PCL-718 from Labcard [4]);
- a personal computer.

The equipment is completed with an oscilloscope. Two additional options are considered:

- eccentric gear connection to allow simulation of the wheel wear mentioned above which can be implemented by either an eccentric wheel or an unaligned axis;
- a signal generator to add extra signals at different points in the loop to simulate periodic disturbances and/or noise;
- optionally, a CADCS software package such as PC-MATLAB [5] or a high-level language with easy access to program the low-level interfaces.

In both cases, the basic set-up depicted in Figure 8.7, relative to the connection between the acquisition card used and the d.c. motor, is necessary owing to a card output (D/A) range of $0 - +4\,\mathrm{V}$.

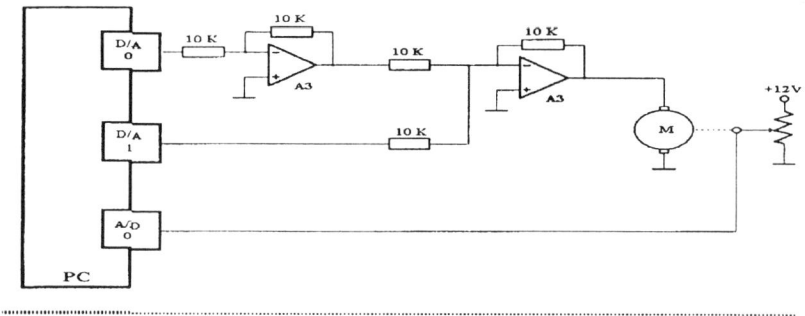

Figure 8.7   Connection between PCL-718 Labcard and d.c. motor

8.4.1   Model

The mathematical description of the process is a continuous time, second-order model, where the output $y(t)$ represents the axis position, $d/dt[y(t)]$ the angular speed and $u(t)$ is the input voltage.

With no load, the experimental parameter identification procedure leads to the following continuous transfer function description:

$$G(s) = \frac{Y(s)}{U(s)} = \frac{1022.85}{s(s+73.6)}$$

As explained later, a sampling period of $T = 0.05$ s has been selected. The discrete state-space description is:

$$x[(k+1)T] = Ax(kT) + Bu(kT); \qquad y(kT) = Cx(kT)$$

where

$$A = \begin{bmatrix} 1 & 0.0132 \\ 0 & 0.0252 \end{bmatrix}; \quad B = \begin{bmatrix} 0.5108 \\ 13.5469 \end{bmatrix}; \quad C = \begin{bmatrix} 1 & 0 \end{bmatrix}$$

A d.c. motor dead zone was detected.

8.5 Suggested experiments

The plant allows the testing of most of the concepts mentioned above:

- low frequency disturbance reduction — a PID controller will reject the static and low frequency components of the load;
- measurement noise filtering;
- disturbance detection and counteraction.

8.5.1 Basic controller

Let us assume the required closed-loop reference response specifications to be an overshoot of $\delta \leq 4\%$ and a settling time $t_s \leq 1$ s. Owing to the requirement for constant speed feeding, an integral term is needed on the controller, so a PID control was designed by means of the well known root locus method. The result was

$$G_R(s) = K\left[1 + T_d s + \frac{1}{T_i s}\right]$$

where $K = 0.675$, $T_d = 0.0056$ s, and $T_i = 2.25$ s. To select the sampling time, the d.c. motor output transient behaviour and some operational constraints were taken into account:

- the shaft's final position should achieve steady state in a smooth way;
- the control actions must be lower than the range of saturation of the A/D channels.

With these conditions taken into account, a sampling period $T = 0.05$ s was chosen. The discrete controller can be obtained by using the expressions [2]:

$$q_0 = K \times (1 + T_d/T) = 0.75$$
$$q_1 = -K \times (1 + 2T_d/T - T/T_i) = -0.81$$
$$q_2 = K \times T_d/T = 0.075$$

8.5.1.1   *Disturbance analysis*

The periodic disturbance is injected into the closed-loop system by means of a signal generator at the input level.

We consider a periodic disturbance $d(t) = d(0) \times \cos wt$, where w represents the frequency. The disturbance state-space representation is:

$$\frac{d}{dt}[x_d(t)] = \begin{bmatrix} 0 & 1 \\ -w^2 & 0 \end{bmatrix} \times x_d(t); \qquad d(t) = \begin{bmatrix} 1 & 0 \end{bmatrix} \times x_d(t)$$

The discrete model obtained for $T = 0.05$ s and $w = 8$ rad/s is

$$x_d[(k+1)T] = A_d \times x_d(kT); \qquad d(kT) = C_d \times x_d(kT)$$

where:

$$A_d = \begin{bmatrix} 0.921 & 0.3894 \\ -0.3894 & 0.921 \end{bmatrix}; \qquad C_d = \begin{bmatrix} 0 & 0.2810 \end{bmatrix}$$

The theoretical step-reference response of the controlled process is shown in Figure 8.8.

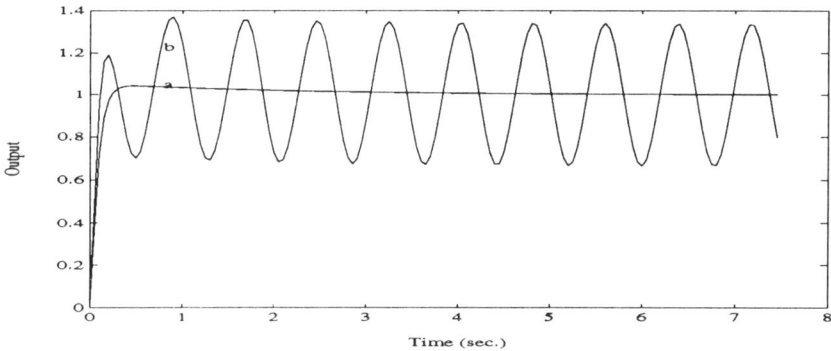

Figure 8.8　　*Theoretical step-reference response*

In response (a) no disturbance is considered, whereas response (b) shows the effect of the periodic disturbance.

8.5.2　Feedback filter

A potentiometer generates a noisy positional measurement. This disturbance is easily removed by filtering the measurement device output. On the other hand, to cancel the $w = 8$ rad/s output component, a notch filter has been designed [6]:

$$\overline{F}(s) = \frac{s^4 + 40s^2 + 400}{s^4 + 26.9s^3 + 401s^2 + 537.4s + 400}$$

This filter is only of academic interest, as the complexity involved results in a filter that is difficult to tune and has numerical problems in the equivalent feedback filter discrete time transfer function. For this case the following was obtained:

$$F(z) = \frac{7.36z^8 - 18.76z^7 + 6.88z^6 + 19.16z^5 - 19.32z^4 - 1.55z^3 + 4.98z^2 - 1.95z + 0.1}{1.82z^8 - 5.43z^7 + 2.1z^6 + 8.56z^5 - 9.49z^4 - 0.8z^3 + 5.38z^2 - 2.3z + 0.1823}$$

As expected, the closed-loop reference time response obtained, Figure 8.9, shows poor behaviour, quite different from that required, owing to the effect of the filter.

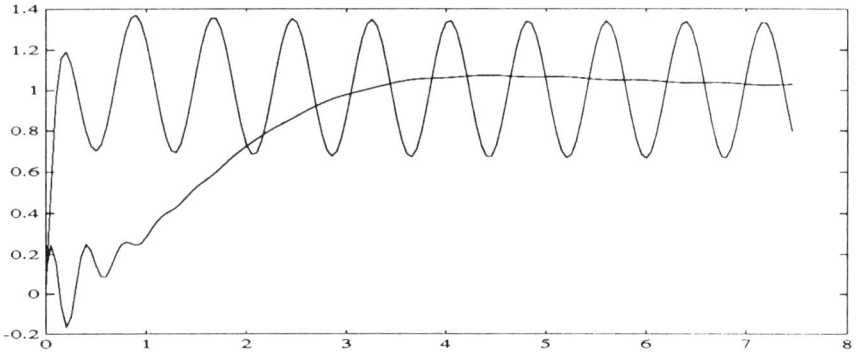

Figure 8.9   Theoretical step-filtered reference response

A frequency domain analysis of the closed-loop transfer function $y(w)/d(w)$, as shown in Figure 8.10, leads to the following conclusions:

- low frequency gain is reduced, due to the integral control action;
- gain at 8 rad/s frequency is reduced due to the notch filter effect, but a range of frequencies is also distorted;
- high frequency disturbances, such as measurement noise, can easily be removed by filtering.

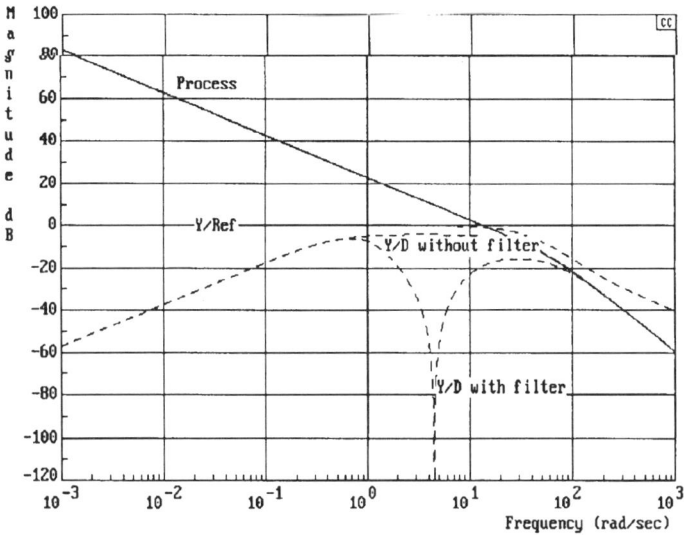

Figure 8.10   Frequency plot of reference and disturbance responses

8.5.3 Disturbance estimator

We now apply the disturbance estimation approach previously outlined. The disturbance observer designed was planned to be four times faster than the closed-loop response. Thus, the discrete characteristic equation for the estimator is:

$$\alpha_{est}(z) = z^2 - 1.6z + 0.67 = 0$$

Where z is the discrete variable. The theoretical result for the assumed case is shown in Figure 8.11a.

As can be seen, a problem was detected in the steady-state estimation of the periodic disturbance, when certain kinds of d.c. motor are used: there is a phase shift. This problem was solved by introducing a phase-lead filter $D(z)$ [6,7], see Figure 8.11b.

$$D(z) = \frac{w_{wp}(w_{wo} + 2/T)}{w_{wo}(w_{wp} + 2/T)} \frac{\left[z + \dfrac{w_{wo} - 2/T}{w_{wo} + 2/T}\right]}{\left[z + \dfrac{w_{wp} - 2/T}{w_{wp} + 2/T}\right]}$$

where

$$w_{wo} = \frac{1}{T}; \quad w_{wp} = \frac{1}{\alpha T}$$

and

$$\sin\theta = \frac{1-\alpha}{1+\alpha}$$

where θ is the phase shift leading to the time response shown in Figure 8.11b.

The closed-loop time response is better and the disturbance effect is almost fully cancelled out. The transient response is satisfactory, and no steady-state degradation has been detected in most cases, but care must be taken under certain conditions, especially if backlash phenomena appear.

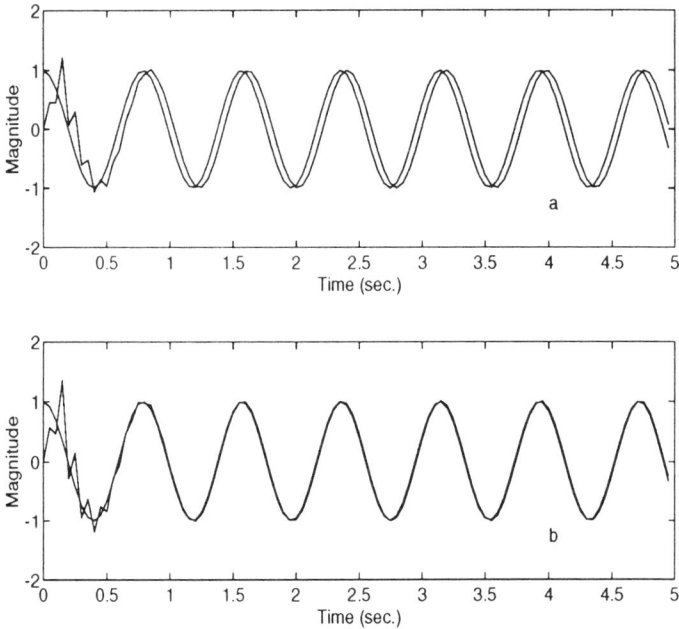

Figure 8.11   Disturbance estimation
a observer output
b phase-lead filter

As mentioned previously, some parameters in the disturbance model may be unknown or time-variant. In this case, a classical parameter estimation algorithm can be implemented allowing tracking of the disturbance model. In the following example two different situations are considered:

(i)   a change in the disturbance frequency appears at a time of 9 s (Figure 8.12) which emulates the case of variation in the speed of the welding operation—the disturbance estimation block follows this change;

(ii)   a change in the disturbance amplitude, reflecting a change in the axis alignment (Figure 8.13). Again, the estimation block works very well.

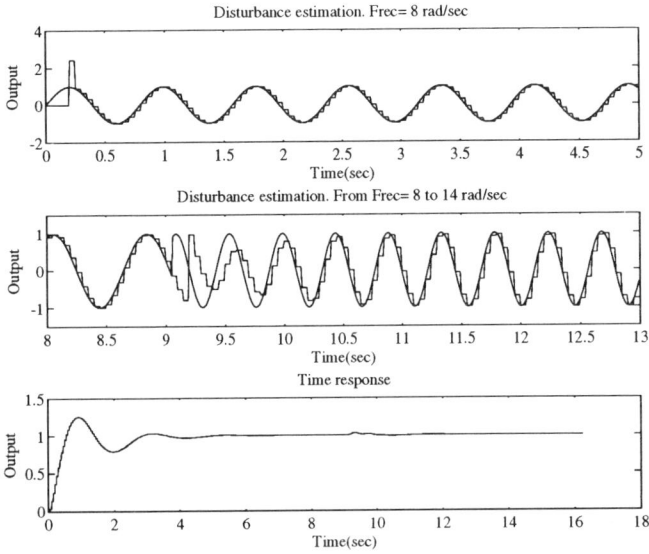

Figure 8.12   Step change in the disturbance frequency

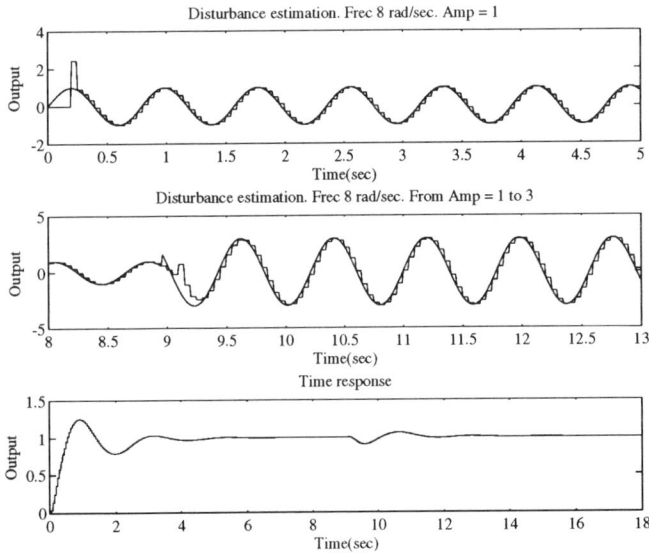

Figure 8.13   Sudden change in the disturbance amplitude

8.6   Illustrative experimental results

In this section, a set of results obtained working with the experimental set-up is shown, which are in agreement with the theoretical simulations given previously.

The first experiment exhibits the disturbance effect: a step change in the reference position is introduced (Figure 8.14).

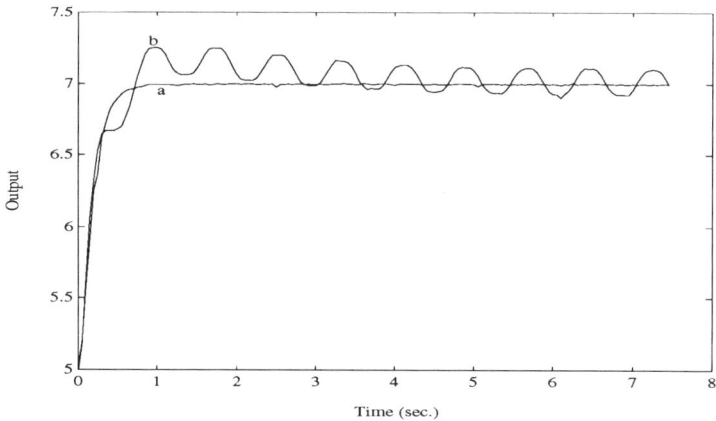

Figure 8.14   Experimental closed-loop step-reference response

The output is analysed in both cases (with and without the disturbance). Case (b) shows the effect of the 8 rad/s periodic disturbance, and curve (a) is the result obtained without the injection of the external disturbance.

The second experiment is carried out to try to counteract the disturbance. The disturbance itself has been estimated and a 'feedforward' control signal has been added to the controller output. The same step response is plotted in Figure 8.15.

In this case, a steady-state sinusoidal disturbance was generated. Figure 8.16 compares the estimated and real disturbances.

8.7   Conclusions

To counteract the undesirable effect of plant disturbances, a number of options were considered. The solutions proposed, although illustrated by a real control problem, are mainly academically oriented, but we try to give guidelines to deal with these kind of problems.

It should be noted that there is always a trade-off between disturbance rejection and reference response. Any controller design based on a set of specifications with only one input response must be carefully checked with respect to other forms of input.

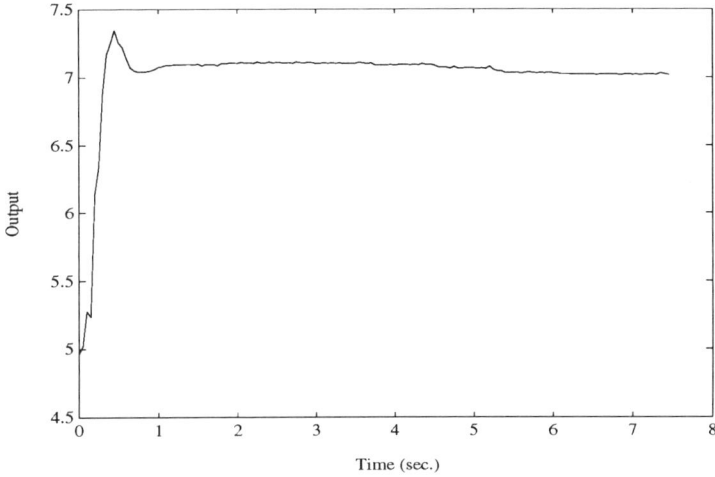

Figure 8.15   Experimental closed-loop reference response with disturbance rejection

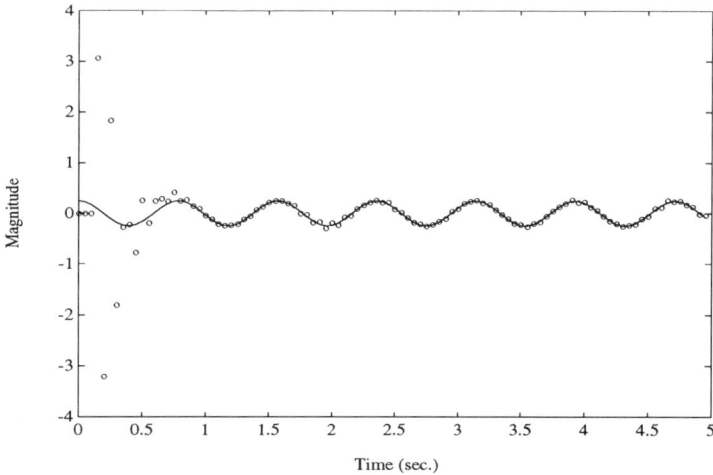

Figure 8.16   Comparison of estimated and real disturbances
o   estimated
—   real

8.8   References

1   CRUZ, J.: 'Feedback systems' (McGraw-Hill, 1972)

2   ISERMANN, R.: 'Digital control systems' (Springer-Verlag, 1989)

3   'Real time control handbook' (Quanser Consulting)

4   'Pc Labcard 718 user's guide' (PcLabcard)

5   'PC-MATLAB user's guide' (The Mathworks Inc.)

6   BOZIC, S. M.: 'Digital and Kalman filtering' (Edward Arnold, 1979)

7   PHILLIPS, C. L. and NAGLE, H. T.: 'Digital control systems: analysis and design' (Prentice Hall, 1984)

Chapter 9

Multivariable process control

N. Mort

9.1   Introduction to multivariable systems

The majority of published work on feedback control systems concentrates on dynamic systems which are single-input, single-output (SISO). However, it is readily acknowledged that many processes found in the process, aerospace and marine industries have more than one input and output and are therefore termed multi-input, multi-output (MIMO) systems. These systems pose a new set of problems when it comes to control because of the effect of one input on more than one output. This can be shown diagramatically in Figure 9.1 where a traditional transfer function representation has been used. Note that the input $R_1(s)$ not only affects the output $Y_1(s)$ according to the transfer function $G_{11}(s)$, but also the output $Y_2(s)$ by the cross-coupling term $G_{12}(s)$. There is a similar effect for the output $R_2(s)$ with respect to the outputs $Y_1(s)$, $Y_2(s)$.

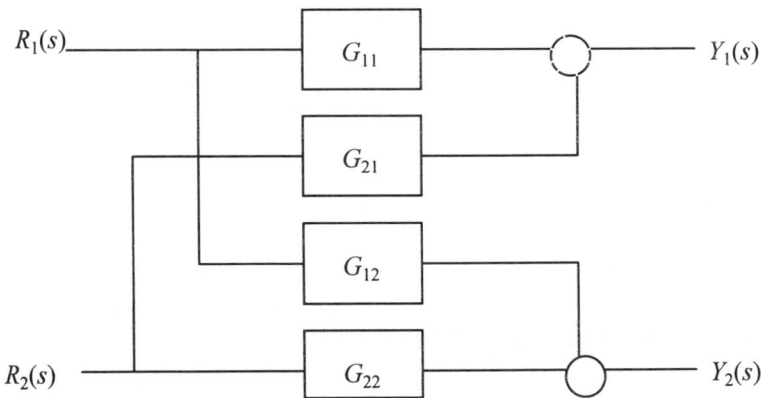

Figure 9.1   Block diagram of a two-input, two-output multivariable system

Considerable effort has been applied to the study of MIMO systems with a significant proportion of the work concentrating on reducing the effect of the cross-coupling terms. Pioneering work in this field can be attributed to Rosenbrock [1,2], Macfarlane [3,4], Owens [5] and others.

MIMO processes are commonplace in industry. For example, in the rolling of strip steel, both speed and tension must be adjusted simultaneously in order to maintain the quality of the finished product. In the chemical process industry, the distillation column presents a good example of a non-linear, time-varying multivariable system. There are also numerous examples in the aerospace and marine industries such as military and civil aircraft, surface ships and submarines. In all these cases, there is more than one control surface/actuator and each actuator can influence a number of controlled variables. There is even a common example of a multivariable system in the home — the domestic electric shower unit. The effect of a cross-coupling term in the operation of a shower can be seen when a tap is opened elsewhere in the house while the shower is running. The flow is reduced and the temperature increases (often reaching unacceptably high levels!) In multivariable terms, we can consider the inputs to the shower as the pump voltage and the heating element voltage. The outputs are the flow rate and temperature of the water.

9.2   Process modelling

9.2.1   Laboratory process: motor-alternator set

Most of the techniques of classical SISO control presume the existence of a suitable mathematical model of the plant to be controlled. Often, the plant is assumed to be linear, time-invariant so that the dynamics may be described by linear differential equations. Clearly, the assumption of linearity cannot be sustained with industrial processes over their complete operating range so it is common to designate an 'operating point' for the specification of a model and then to consider small signal changes from the equilibrium condition.

The above discussion applies equally well to MIMO systems and we can explore techniques of process modelling by means of a laboratory scale process, the motor-alternator set. This is a small-scale model of a turbo-generator system that would be found in an electricity supply power station. The laboratory process consists of a d.c. motor directly coupled to an alternator with both components rigidly mounted on a common steel bed. The voltage levels of the plant variables can be regulated by a manual controller in order to set a specific operating point for the machine. A block diagram of the unit is shown in Figure 9.2.

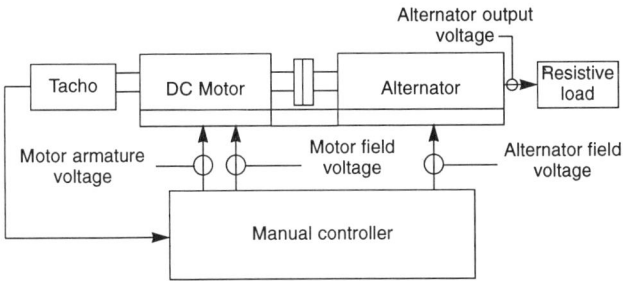

Figure 9.2   Block diagram of motor-alternator laboratory system

This plant can be represented as a MIMO system with two inputs and two outputs:

- alternator field voltage u_1

- alternator field voltage u_2

- alternator output voltage y_1

- alternator output frequency (motor speed) y_2

Thus it is possible to consider this process with the structure shown earlier in Figure 9.1 and the modelling task becomes a requirement to identify the transfer functions G_{11} G_{12} G_{21} and G_{22}. It is common to represent this structure in vector/matrix form as:

$$\begin{bmatrix} y_1 \\ y_2 \end{bmatrix} = \begin{bmatrix} G_{11} & G_{12} \\ G_{21} & G_{22} \end{bmatrix} \begin{bmatrix} u_1 \\ u_2 \end{bmatrix}$$

(9.1)

9.2.2   Model identification tests

It is often difficult to determine a model of a practical system from an analysis of the physical relationships within the process. An alternative is to apply test signals to the designated input(s) of the process, measure the response at the output(s) and deduce the parameters of the model by analysing the input/output data. These techniques of 'systems identification' or 'parameter estimation' are well documented in the literature, e.g. References 6 and 7 and Chapter 1. Some of the techniques rely on special input signals such as PRBS (pseudo random binary sequences) but a simpler approach to give a first approximation of the process is to use step input test stimulation. The advantages of step testing are:

(a)   Step tests are simple to perform and no special equipment is needed;
(b)   Experimental time for one response is approximately five times the dominant time constant of the plant;
(c)   The method is good for gain estimates and adequate dynamic response characteristics, especially if the process is of a low order (first or second).

For the motor alternator the procedure was to apply a unit voltage step to u_1 while u_2 is held constant and record the output transient responses in y_1 and y_2 then to repeat for u_2 with u_1 held constant. The recorded data is then processed using appropriate software; typically MATLAB might be used for displaying the results on a VDU after mean levels had been removed. The results for these tests are shown in Figures 9.3 and 9.4.

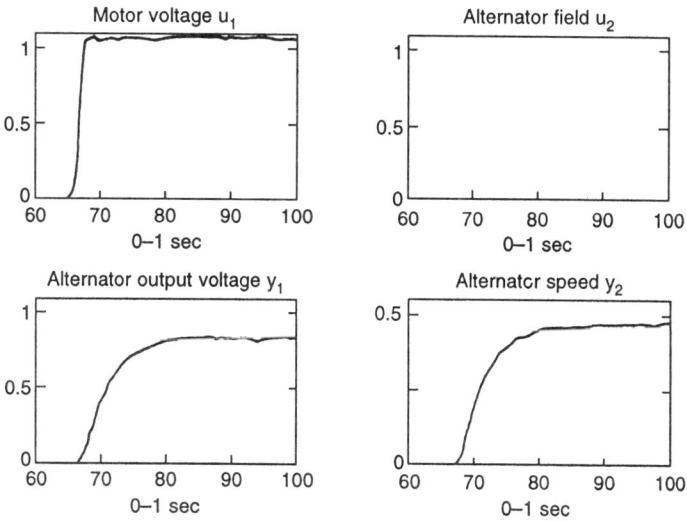

Figure 9.3   *Responses due to step change in motor voltage u_1*

From these figures, it can be seen that the responses are approximately first order. Further examination to assess the gain and time constants reveals that an approximate mathematical model of the plant can be written as

$$Gp = \begin{bmatrix} \dfrac{6.1}{s+7} & \dfrac{4.6}{s+14.3} \\ \dfrac{2.8}{s+6.8} & \dfrac{-0.24}{s+4} \end{bmatrix} \qquad (9.2)$$

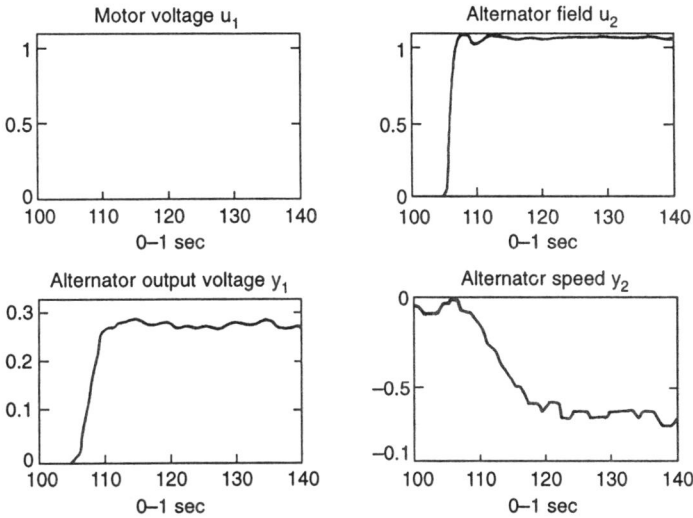

Figure 9.4   *Responses due to step change in alternator field voltage u₂*

9.3   Multivariable controller design

It has been stated earlier that one of the main difficulties in designing controllers for multivariable systems is the presence of the cross-coupling terms. A substantial amount of effort has been expended in attempting to understand the effects of the interactions. A useful summary of these methods is contained in Sinha [9].

In this section we will illustrate two possibilities that could be used to deal with the typical plant considered in the last section, the motor-alternator unit. In both cases, the design of a pre-compensator is involved which is intended to address the problem.

9.3.1   Non-interacting control

Theory
This design methodology is relatively simple to apply but it does impose severe constraints on the pre-compensator. Consider the simple block diagram structure shown in Figure 9.5

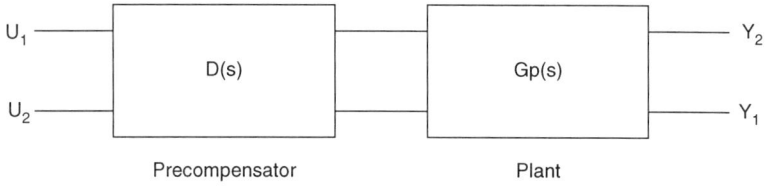

Figure 9.5   Pre-compensator and multivariable plant

The basic idea can be stated as:

(1) Let $Gp(s)$, the known plant transfer function matrix, be written as:

$$Gp(s) = R(s).A(s)$$

where:
   $R(s)$ is a diagonal matrix containing the common row elements of $Gp(s)$

   $A(s)$ is a matrix of numerator terms only

(2) Choose $D(s) = A^{-1}(s)$

(3) Then $Gp(s).D(s) = R(s)$, a diagonal matrix

Thus, the combined system, plant and pre-compensator, is non-interacting.
   It should be noted that the inclusion of $D(s)$ degrades both transient and steady-state performance of the process and additional corrective action would be necessary to restore performance to its original level.

Design for the motor-alternator set
The transfer function matrix has been derived from experimental tests — see Equation 9.2.
   We must represent $Gp(s)$ as $R(s).A(s)$ where these have been defined earlier, i.e:

$$\begin{bmatrix} g_{11} & g_{12} \\ g_{21} & g_{22} \end{bmatrix} = \begin{bmatrix} r_{11} & 0 \\ 0 & r_{22} \end{bmatrix} \begin{bmatrix} a_{11} & a_{12} \\ a_{21} & a_{22} \end{bmatrix} \tag{9.3}$$

By appropriate algebraic manipulation, we deduce that

$$A(s) = \begin{bmatrix} 6.1(s+14.3) & 4.6(s+7) \\ 2.8(s+4) & -0.24(s+6.8) \end{bmatrix} \tag{9.4}$$

and therefore:

$$A^{-1}(s) = \frac{1}{12.88(s^2 + 11s + 28)}\begin{bmatrix} 0.24(s+6.8) & 4.6(s+7) \\ 2.8(s+4) & -6.1(s+14.3) \end{bmatrix} \quad (9.5)$$

But $D(s)$, the decoupling pre-compensator, is $A^{-1}(s)$, so the design is complete.
    Simple matrix arithmetic will show that the product $Gp(s).D(s)$ is, in fact, $R(s)$, a diagonal matrix. Thus, two independent S.I.S.O. control loops can be designed.

9.3.2   The characteristic locus method

Theory
This method of dealing with interaction (see Kouvaritakis [10]) has been selected because of its close relationship with ideas of established classical control theory. The characteristic loci of a matrix $Q(s)$ are the set of loci in the complex plane traced out by the eigenvalues of $Q(s)$ as s traverses the standard Nyquist contour in a clockwise direction. In summary, the method consists of designing a series of cascaded compensators, each of which attempts to deal with a different part of the frequency spectrum, i.e. low, middle and high. The steps to be followed in this design procedure can be listed as:

(1)   Compute a real decoupling compensator $K_h \approx G^{-1}(j\omega_b)$ where ω_b is the bandwidth of the open loop system;

(2)   Design an approximate commutative controller $K_m(s)$ at some frequency $\omega_m < \omega_b$ for the compensated plant $G(s).K_h$ such that $K_m(j\omega) \Rightarrow I$ as $\omega \Rightarrow \infty$ (the latter requirement is an attempt to ensure that the decoupling effected by K_h is not disturbed too much). Ideally, we would like $K_m(j\omega)$ to approach I as $\omega \Rightarrow \omega_b$ but this is not realistic in practice.

(3)   If the low frequency behaviour is unsatisfactory (typically because of excessive steady-state errors), design an appropriate commutative controller $K_1(s)$ such that $K_1(j\omega) \Rightarrow I$ as $\omega \Rightarrow \infty$.

(4)   Realise the complete compensator $K(s) = K_h.K_m(s).K_1(s)$

The subscripts h, m, l denote high, medium and low frequency, respectively. Usually K_1 is used to introduce integral action.

Design for the motor-alternator set

(a) Performance specification

To begin the design of a suitable compensator for the laboratory motor-alternator set, it was necessary to have a specification detailing the performance requirements for the compensated system, $G(s).K(s)$. The proposed specification is outlined below in Table 9.1

System parameter	Required tolerance
Steady state error	±2%
Overshoot	±25%
Settling time	0.5 s
Bandwidth	$60 \, \text{rad s}^{-1} \approx 10 \, \text{Hz}$
Interaction	5%

Table 9.1 Performance specification

(b) Justification for compensation

To verify the requirement for compensation, and to illustrate the time domain performance of the uncompensated system, the closed-loop unit step response for the model can be simulated using, for example, MATLAB. As an illustration, the unity feedback for a unit step change in input 1 (motor armature voltage) is shown in Figure 9.6.

It can be seen that in addition to the interaction, there is a large steady state offset and the performance specification given in Table 9.1 is not satisfied.

(c) Design procedure

The first step is to calculate the multivariable frequency response (MVFR) matrix for the uncompensated system over a suitable frequency range $(0.1 \rightarrow 100 \, \text{rad s}^{-1})$. It is then possible to obtain a graphical representation of the characteristic loci by calculating the eigenvalues of the MVFR matrix. This computationally-intensive operation can be carried out easily using the multivariable frequency domain (MFD) toolbox in MATLAB. This toolbox provides an invaluable extension to the standard functions in MATLAB and is an excellent aid in the design process for multivariable controllers.

(d) Scaling compensator

The Bode magnitude array for the uncompensated motor-alternator (Figure 9.7) clearly shows that the elements of column 2 are an order of magnitude smaller than those of column 1. It proves necessary to balance this difference with a constant gain matrix compensator K_1 to make subsequent design stages easier. Here, we choose K_1 to be:

$$K_1 = \begin{bmatrix} 1 & 0 \\ 0 & 10 \end{bmatrix}$$

and use the appropriate function in MFD to carry out the pre-multiplication.

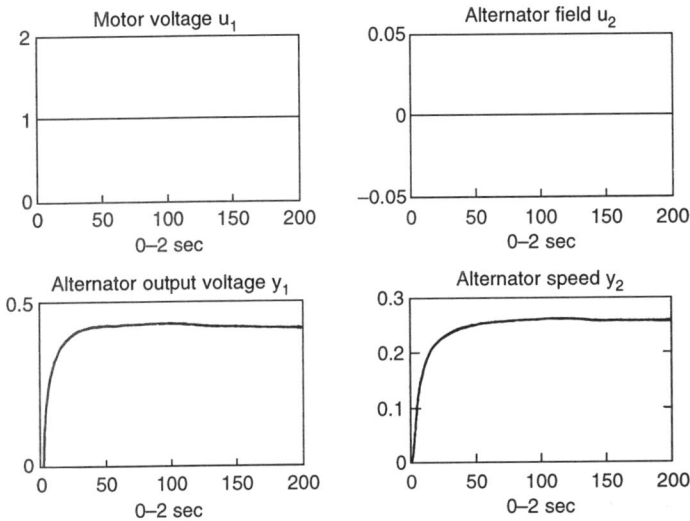

Figure 9.6   Closed loop response of the uncompensated system

(e) Decoupling pre-compensator

The purpose of this compensator is to decouple the interaction at frequencies outside the desired operating bandwidth, so it is usual to implement the design at or near ω_b. After some trial design, the final form of K_h was chosen to be

$$K_h = \begin{bmatrix} 0.64 & 2.35 \\ 0.3 & -0.23 \end{bmatrix}$$

(f) Approximate commutative controller

The purpose of designing a mid-frequency compensator is to decrease the response time and settling time by increasing the gain over the bandwidth, while ensuring that gains outside the bandwidth are low to protect the noise rejection properties of the system. As with most designs, the final proposal reduces to a compromise between the conflicting requirements of a wide bandwidth for fast response and settling time and a narrow bandwidth for adequate noise rejection. One way of achieving this compromise is via a lead-lag network which ensures gain increases occur within the system bandwidth. Thus, K_m is selected to have the structure:

$$K_m = \frac{(1+\tau_1 s)(1+c\tau_2 s)}{(1+\tau_2 s)(1+c\tau_1 s)}$$

where $\tau_1 > \tau_2 > c\tau_1 > c\tau_2$

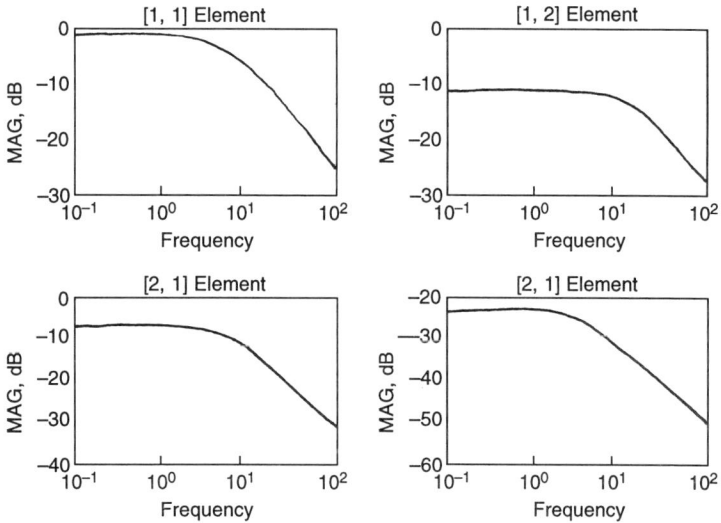

Figure 9.7   Bode magnitude array for the uncompensated system

(g) Low-frequency compensator

The function of this compensator is to provide integral control action to eliminate steady state offset in the system feedback loops and also to increase the low-frequency gains, thus reducing low-frequency interactions but without decreasing the stability margins.

Controller performance

The complete controller consists of the combination of the individual compensators considered in (d), (e), (f), and (g) above. The operations described in the development of the compensators can be largely automated using the MFD Toolbox in MATLAB.

To test the performance of the compensated design, simulation of plant plus compensator can be readily carried out, again using MATLAB. The structure of the closed loop system with compensation is shown in Figure 9.8.

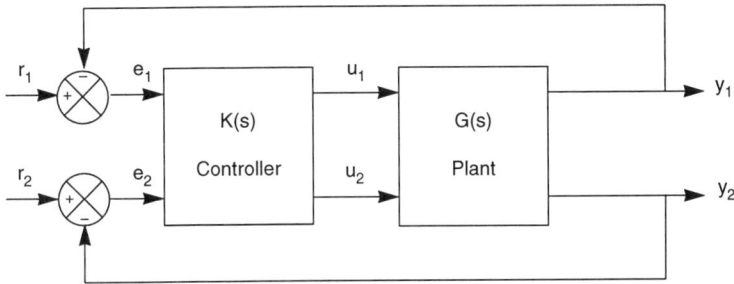

Figure 9.8   *Closed loop compensated motor-alternator system*

The response of both outputs y_1, y_2 for a unit step demand in r_1 are shown in Figure 9.9. It can be clearly seen that the controller meets the requirements of the specification with only a transient interaction in the speed response, output y_2.

The responses shown in Figure 9.9 are simulation results derived from a MATLAB model of the motor-alternator together with the compensator designed using the characteristic locus method. The use of computer-aided techniques has therefore assisted in both the design phase and the testing phase of the overall compensator derivation. What remains to be done, of course, is the final implementation of the compensator on the actual process in the laboratory. There are additional issues to be considered in any practical implementation and we note some of these in the next section. It is sufficient here to state that the implementation of the compensator design can, in principle, be achieved either by analogue or digital methods. If a suitable PC is available with the necessary A/D and D/A interface card, then the digital implementation would be a natural way forward. The procedure to follow involves conversion of the transfer function matrix description of the compensator into the equivalent state space form and then derivation of the recursive equations which approximate the continuous time integration of the state differential equations.

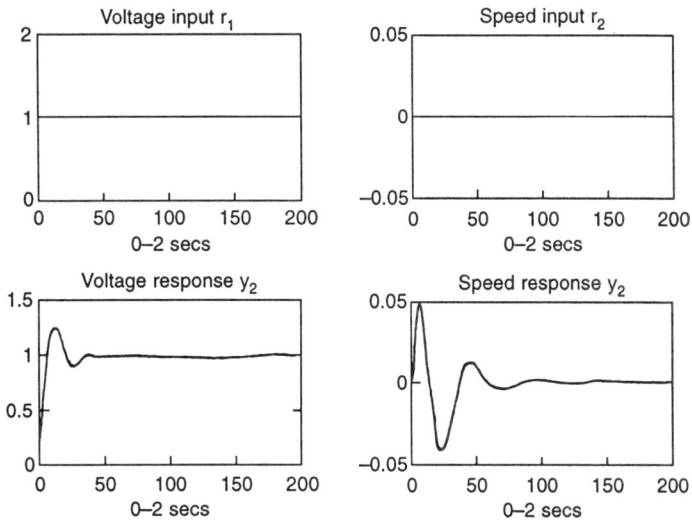

Figure 9.9   Closed loop response for step input at r₁

9.4   Additional experimental work

9.4.1   *Implementation of a digital compensator for the motor-alternator*

(a) In the preceding section, we drew attention to the possibility of writing the necessary code on a PC to represent the compensator's dynamic behaviour. In any real process, there are new problems that must be addressed which are not present in a pure simulation study. For example, signals which need to be monitored on the motor-alternator rig are invariably noisy and special care must be taken in the design process.

The compensator design should be as insensitive as possible to the dominant frequencies of the noise present in the recorded voltage. A very unsatisfactory situation will occur if the loop gain of the feedback system is significant at the high-frequency end of the range. In addition, it is useful to consider the application of suitable filtering techniques to remove or reduce the unwanted noise, as already discussed in chapter 8. The frequency content of the signals should first be checked to determine the dominant components. The peaks in the power spectral density (PSD) function of the signal will indicate these. Having identified the dominant frequencies, a digital filter with a suitable cut-off frequency can be designed and applied to the recorded data. Once again, there are facilities within MATLAB to determine the PSD and the parameters of an appropriate digital filter.

(b) Non-linearities are another factor that begins to influence any real implementation of a digital controller. All the theoretical ideas that have been outlined thus far make the assumption of linearity in the plant parameters. If we restrict our attention to small signal deviations from an operating point then this assumption is generally valid. It is possible, of course, that the system may be subjected to large disturbances and the assumption of linearity is then somewhat suspect. The design of controllers for non-linear systems is beyond the scope of this chapter but there is considerable opportunity for further investigation in this area.

(c) The effect of a variation in the sampling rate in the digital compensator is an additional 'degree of freedom' that is present in real computer control systems.

9.4.2   An alternative multivariable process

It has been shown that the concepts of multivariable control can be applied to laboratory-scale equipment, i.e. the motor-alternator set. Indeed, this equipment can be constructed with minimal cost outlay using, for example, used automobile spares (car dynamo and alternator). The dynamics of the motor-alternator are relatively fast and it must be acknowledged that there are industrial processes with much longer time constants. Typical examples are found in thermal processes, e.g. furnaces, kilns etc. These slower dynamic systems often display pure time delays in the response and these pose an additional difficulty in the design of control systems, as treated in chapter 6.

We can now briefly consider a different laboratory-scale process which is multivariable in nature but has a completely different response time scale. The equipment shown in Figure 9.10 is a process control rig in which there are two variables to control, namely flow rate of water and water temperature. The inputs or 'actuators' of this system are a pump and a heating element, both of which are electrically operated.

The dynamic behaviour of this apparatus can be determined using step testing in the same way as for the motor-alternator equipment. Particular features that can be explored using this rig are:

- Effect of time constants in the model which differ by an order of magnitude. For this process, the temperature response has a time constant of approximately 50 s while the flow rate response time constant is only 0.1 s.

- Effect of time-varying parameters such as the reservoir temperature of the water. If the equipment is operated continuously for some period of time, then the reservoir water temperature increases and this will clearly influence the behaviour of any closed-loop control system for water temperature.

- Effect of process time delays, which are particularly noticeable in the case of the temperature loops.

- Effect of different magnitudes of step input on the estimated parameters of the model. It can be shown relatively easily that the gain estimates for a 5% change from an operating point will be different to those obtained for a 20% change. This effect demonstrates the non-linear nature of many real processes in their dynamic response characteristics.

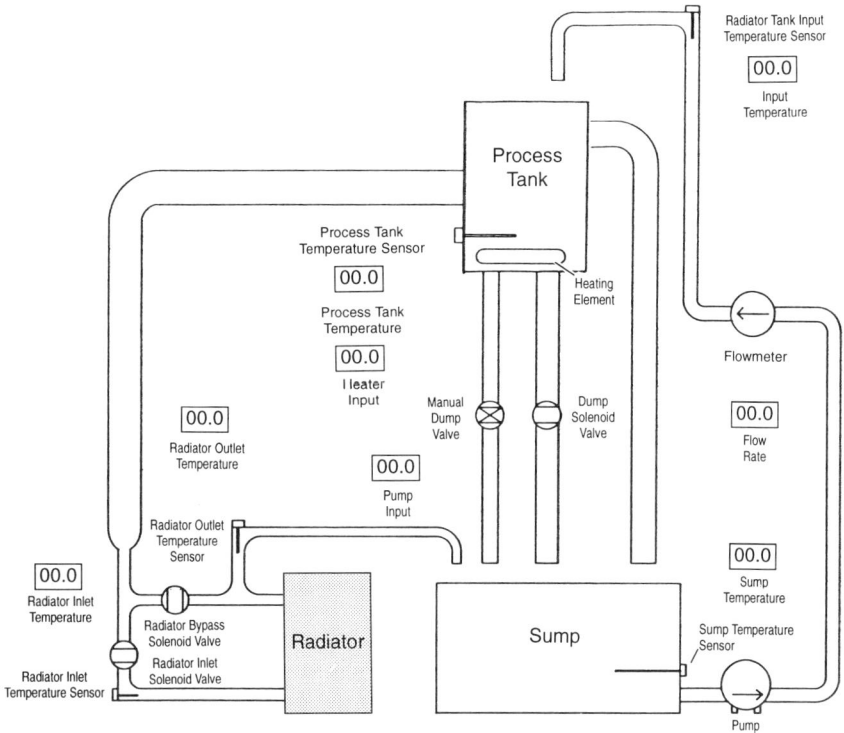

Figure 9.10   The process control apparatus

9.5   Summary

This chapter has introduced the idea of a multivariable system and an example of a laboratory-scale model of such a process has been described. The presence of interactions in the process has been considered and two methods for dealing with these effects have been explored. The use of MATLAB for both simulation study and multivariable controller design has been pointed out and implementation issues have been addressed. The material here forms the basis for further study of multivariable systems and has the added attraction of highlighting real laboratory analogues of common industrial processes that are multi-input and multi-output. For further information on the technical details of the experimental rigs described in this chapter, see Thomas [11] and Burney [12], respectively. More details concerning the theory and practice of multivariable systems can be found in Maciejowski [13], Skogestad and Postlethwaite [14] and Deshpande [15], for example.

9.6   References

1   ROSENBROCK, H. H.: 'Progress in the design of multivariable control systems', *Meas. Control*, 1971, **4**, pp.9-11

2   ROSENBROCK, H. H.: 'Computer-Aided Control Systems Design', Academic Press, London, 1974

3   MACFARLANE, A. G. J. and KOUVARITAKIS, B. A. 'A design technique for linear multivariable feedback systems', *Int. J. Control*, 1977 **25**, pp.837-874

4   MACFARLANE A. G. J. and POSTLETHWAITE I.: 'The generalised Nyquist stability criterion and multivariable root-loci', *Int. J. Control*, 1977, **25**, pp.81-127

5   OWENS D. H.: 'Dyadic expansion for the analysis of linear multivariable systems', *Proc. IEE*, **121**, 1974, pp.713-716

6   LJEUNG, L. J.: 'System Identification', Prentice-Hall, London, 1987

7   BILLINGS, S. A. and CHEN, S.: 'Neural Networks for nonlinear dynamic system modelling and identification', *Int. J. Control*, 1992, **56**, pp.319-346

8   MATLAB Users Manual, The Mathworks Inc.

9   SINHA, P. K., 'Multivariable Control - An Introduction', Dekker, New York, 1984

10   KOUVARITAKIS, B. A.: 'Theory and practice of the characteristic locus design method', *Proc. IEE*, 1979, **126**, pp.542-548

11   THOMAS, M. A.: 'The Design of a Multivariable Controller for a Model-Scale Motor/Generator Set', Project Dissertation, Dept. of Automatic Control & Systems Engineering, University of Sheffield, 1992

12   BURNEY, P.: 'Modelling of a Temperature and Flow Process Control Rig', Project Dissertation, Dept. of Automatic Control & Systems Engineering, University of Sheffield, 1993

13   MACIEJOWSKI, J. M.: 'Multivariable Feedback Design', Addison-Wesley, 1989

14   SKOGESTAD, S. and POSTLETHWAITE, I.: 'Multivariable Feedback Control: analysis and design', Wiley, 1996

15   DESPHANDE, P. B.: 'Multivariable Process Control', Instrument Society of America, 1989

Chapter 10

Predictive control vs. PID control of thermal treatment processes

M. Voicu, C. Lazãr, F. Schönberger, O. Pãstravanu and S. Ifrim

10.1 Introduction

The control of thermal processes has been given much attention over recent years and the use of digital controllers has been suggested for such real-life applications. These controllers often consist of discrete time versions of classical PID controllers or sometimes modern predictive and adaptive controllers. In this chapter a comparison is made between predictive and PID control, exemplified on a practical control problem.

10.2 Control problem

We are considering resistance furnaces for thermal treatments and we want to design a control law so that the furnace temperature changes according to a desired profile imposed by a technologist. A classical thermal cycle with different plateaux and ramps is depicted in Figure 10.1 as our desired profile.

The following problems must be considered when controlling the thermal treatment furnace:

— the desired profile of the regulated temperature is produced by step and ramp functions;
— such plants have time delay;
— the steady-state error must be zero;
— the quality of the thermal treatment process is dependent on the accuracy of the control system.

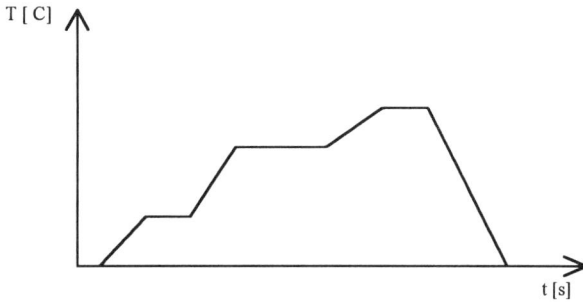

Figure 10.1   Typical desired temperature profile

A furnace with a single temperature zone is usually modelled using a first (or second) order element with a time delay. Estimation of the model parameters can be achieved using the MATLAB System Identification Toolbox facilities and a deterministic auto-regressive moving average (DARMA) model can be obtained.

10.3   Technical approaches to control the thermal treatment processes

Since thermal treatment processes can be accurately modelled by a first-order transfer function with time delay, PID algorithms can sometimes be implemented successfully for controlling such plants, but the time delay has an unfavourable effect on performance. More accurate regulation can be obtained with predictive algorithms.

10.3.1 PID algorithm

Temperature control in most thermal treatment processes has normally been accomplished using PID controllers [1,2]. Considering a linear model (estimated off-line) of the first order plus time delay type:

$$G(s) = \frac{K_f}{T_f s + 1} e^{-\tau s} \tag{10.1}$$

there are many methods for tuning PID parameters [3–6] depending on the form of the controller and the performance criteria, (see also chapter 6).

In many practical situations, derivative action is introduced in the feedback path and is filtered by a first-order lag which reduces the noise effects. In this particular form of the PID control law, the control signal $u(t)$ is given by

$$U(s) = K_p\left(1 + \frac{1}{T_i s}\right) E(s) - \frac{T_d s}{\left(1 + \frac{s T_d}{N}\right)} Y(s) \tag{10.2}$$

where $e(t)$ is the regulation error between the reference signal $y^*(t)$ and the measured output of the plant $y(t)$. The parameters K_p, T_i and T_d of the PID controller (assuming $N=10$) are determined according to the plant coefficients K_f, T_f and τ to minimise the integral of the time and square of error (ITSE) criterion:

$$J_1 = \int_0^\infty t e^2(t) dt \tag{10.3}$$

using the tuning formulas given in Reference 5.

The controller model in the Z domain is obtained, for instance, by using the Euler approximation:

$$U(z) = G_1(z) E(z) - G_2(z) Y(z) \tag{10.4}$$

with:

$$G_1(z) = \frac{K_p[(1 + T/T_i)z - 1]}{z - 1} \tag{10.5}$$

$$G_2(z) = \frac{T_d(z-1)}{(T + 0.1 T_d)z - 0.1 T_d} \tag{10.6}$$

where T denotes the sampling period.

10.3.2 Predictive algorithm

Unfortunately, a time delay has unfavourable effects on the performance obtained with PID control. To eliminate these, an alternative weighted one-step-ahead control is proposed. In Section 10.2 we showed that we obtained by identification a time delay plant that can be described by a DARMA model in the language of Goodwin and Sin [7] that is expressed as:

$$A(q^{-1})y(t) = B(q^{-1})u(t) \tag{10.7}$$

where:

$$A(q^{-1}) = 1 + a_1 q^{-1} \tag{10.8}$$

$$B(q^{-1}) = q^{-d} B^1(q^{-1}) \qquad (10.9)$$

and q^{-1} is the backward shift operator and dT is the time delay.

The DARMA model mentioned above can be expressed in d-step-ahead predictor form as:

$$y(t + d) = E(q^{-1})y(t) + F(q^{-1})u(t) \qquad (10.10)$$

where:

$$E(q^{-1}) = e_0 \qquad (10.11)$$

$$F(q^{-1}) = f_0 + f_1 q^{-1} + ... + f_{d-1} q^{-(d-1)} \qquad (10.12)$$

The coefficients e_0 and f_i, $i = 0,...,d-1$ are determined according to [7].

For the plant described by Equation 10.7, the weighted one-step-ahead control brings $y(t+d)$ to the desired value $y^*(t+d)$ in one step. This control procedure that compensates the time delay may be applied because the future desired output is known from the thermal cycle prescribed by the user.

The control law has the form:

$$u(t) = \frac{f_0[y^*(t+d) - E(q^{-1})y(t) - F^1(q^{-1})u(t-1)] + \lambda P^1(q^{-1})\overline{u(t-1)} - \lambda R^1(q^{-1})u(t-1)}{f_0^2 + \lambda}$$
$$(10.13)$$

where:

$$F^1(q^{-1}) = q[F(q^{-1}) - f_0] \qquad (10.14)$$

$$P(q^{-1})\overline{u(t)} = R(q^{-1})u(t) \qquad (10.15)$$

$$P^1(q^{-1}) = q[P(q^{-1}) - 1] \qquad (10.16)$$

$$R^1(q^{-1}) = q[R(q^{-1}) - 1] \qquad (10.17)$$

and because we chose $\overline{u(t)} = u(t) - u(t-1)$

$$P^1(q^{-1}) = 0; \qquad R^1(q^{-1}) = -1 \qquad (10.18)$$

The control law in Equation 10.13 minimises the following cost function:

$$J(t+d) = \left\{ \frac{1}{2} \left[y(t+d) - y*(t+d) \right]^2 + \frac{\lambda}{2} \overline{u(t)}^2 \right\} \tag{10.19}$$

and because $(1-q^{-1})$ is a factor of $R(q^{-1})$, zero steady-state tracking error is achieved for a constant $y*(t)$ sequence. Thus, a 2-DOF (degree of freedom) controller equipped with integral action is obtained.

By recoding the control law of Equation 10.13 in terms of discrete transfer functions, the predictive control can be expressed as:

$$U(z) = G_1(z)Y*(z) - G_{2(z)}Y(z) \tag{10.20}$$

$$G_1(z) = \frac{f_0 z^{d+1}}{(f_0^2 + \lambda)z + \lambda f_0 F^1(z^{-1}) - \lambda} \tag{10.21}$$

$$G_2(z) = \frac{f_0 E(z^{-1})z}{(f_0^2 + \lambda)z + \lambda f_0 F^1(z^{-1}) - \lambda} \tag{10.22}$$

10.4 Discussion

By adopting appropriate tuning, PID controllers can give results which compare quite well with advanced controllers. The control objective is to guarantee the fastest time response compatible with a small amount of overshoot and no steady-state offset, accounting for the presence of time delay. The most commonly used tuning criteria in industry have been ISE (integral of square error), IAE (integral of absolute error) and ITSE. The most appropriate criterion which partially solves the control problem outlined is ITSE. Unfortunately, time delay compensation and steady-state offset for ramp inputs remain unsolved. The adoption of a PID controller causes an output response which will differ from the desired one $y*(t)$.

This might be obtained exactly by a more complex controller structure given by a weighted one-step-ahead control law. Using a predictive procedure, the time delay can be compensated and the accuracy of the controlled system will be improved.

10.5 Laboratory set-up

The basic elements of the laboratory set-up [8] are:

— resistance furnace for thermal treatment processes;
— commercial data acquisition card;
— personal computer.

The layout of the furnace, including sensor and actuator, is depicted in Figure 10.2. The command *u* is applied to the actuator which consists of a variable voltage regulator (VVR). The temperature is measured by means of a thermocouple. The average power given by the resistor R is determined by the phase control of the triac-based static switch TH and the adapter A yields the regulated output *y*. The furnace resistance is supplied from a 220 V AC supply and is able to develop a power of 2 kW. The control variable is a normalised 0–5 V analogue signal corresponding to 0–100% power. The thermocouple, with adapter, delivers a 0–5 V signal, corresponding to a 0–550 °C temperature range, with a maximum resolution of 1 °C.

The personal computer used for the control system hardware is provided with an interface card having 12 bit A/D and D/A converters that make the connection to the plant.

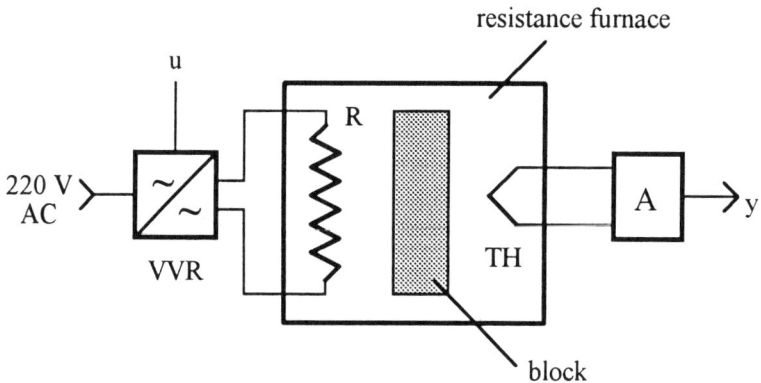

Figure 10.2   Resistance furnace

10.6 Suggested experiments

The following experiments are proposed to study the behaviour of the laboratory set-up and the performance of the PID and predictive control laws, in respect of furnace control.

10.6.1 Parameter identification for the plant

A step response test of the thermal treatment furnace should be performed. On the basis of the set of response values obtained, the physical parameters of the system can be computed using MATLAB Identification Toolbox facilities.

10.6.2 PID control

The behaviour of the PID-controlled loop needs to be simulated to find the optimal tuning parameters of the control law. Real-time experiments can also be performed on the laboratory set-up.

10.6.3 Predictive control

The influence of weighting factor λ in the action of the predictive control law (Equation 10.13) should be studied by simulation. The performance of the predictive-controlled furnace should be tested and compared with that of the PID-controlled loop.

10.6.4 Simulation phase and control algorithm implementation

The simulation phase of the experiments may be carried out using the SIMULINK software package which offers a graphical environment to test the controlled loops described directly in terms of the block diagram of the system.

Unfortunately SIMULINK is restricted to the simulation area and does not allow direct connection to the real plant. To avoid recoding the algorithm to a suitable programming language with I/O facilities and real-time synchronisation, our proposed solution is to provide an additional block library for SIMULINK which allows an I/O interface between the high-level, built-in control tool and the A/D–D/A card (DAC 6214). We prefer to use this solution to shorten the development of the testing process as this method, keeping the task in an integrated environment, ensures easy implementation.

New SIMULINK blocks named C_AD, C_DA and RT_SYNC have been developed to permit connection with the process and synchronisation with a real-time clock during the 'simulation' carried out by SIMULINK.

10.7 Illustrative examples

To test the performance of the PID- and predictive-controlled loops, the experiments suggested in Section 10.5 are conducted under the following conditions:
— sampling period T=120 s;
— the reference signal has an initial step variation followed by a typical thermal cycle with ramps of different positive and negative slopes (bearing in mind physical constraints) and regulations at various levels;
— saturation limits are considered to be 0 and 5 V of the control signal;
— the model structure of the furnace is assumed to be first-order plus time delay type (1) with the coefficients: $K_f = 1.9$, $T_f = 5450$ s, $\tau = 240$ s.

The results (both output and control signals) of the simulation of the PID control loop are shown in Figure 10.3. Notice the regulation error to ramp variations of the reference signal and also the over- or undershoots of the response.

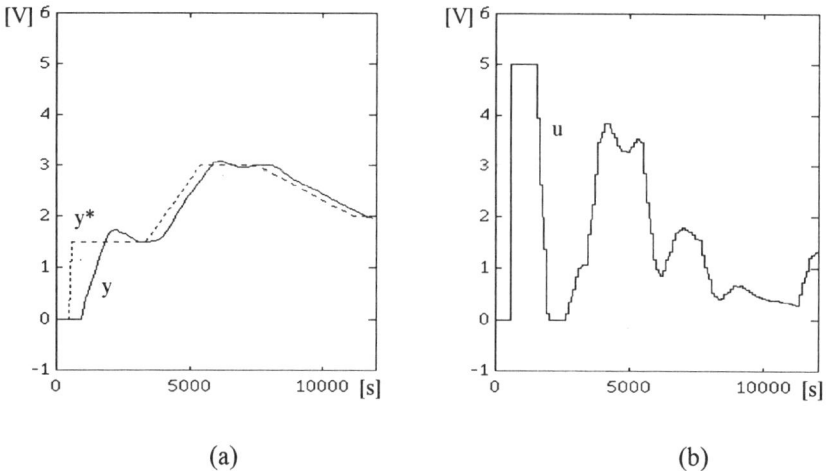

(a)   (b)

Figure 10.3   Simulation results of the PID control loop
(a) closed-loop response
(b) control variable evolution

Better performances are obtained using the predictive algorithm. Figure 10.4 shows that the control objectives are reached, that there is no steady-state error for up and down ramps and over- or undershoots are reduced.

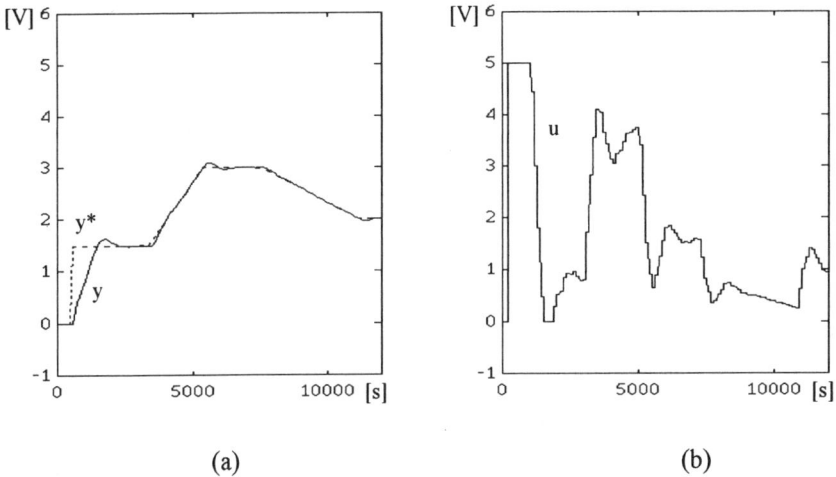

Figure 10.4   *Simulation results of the predictive control loop ($\lambda=0.002$)*
(*a*) closed-loop response
(*b*) control variable evolution

The weighting factor λ is optimally chosen at the value 0.002. If a value of 0.001 is taken for λ (Figure 10.5), the control signal behaviour becomes very erratic, with frequent saturation which can sometimes lead to ON/OFF control during some sample periods. On the other hand, a greater value for the weighting factor λ (towards 0.01, Figure 10.6) makes the control action slow and regulation becomes inaccurate.

Similar performances have been obtained in real-time experiments. Results of the temperature control experiments on the furnace, performed using PID and predictive controllers are reported in Figures 10.7 and 10.8, respectively.

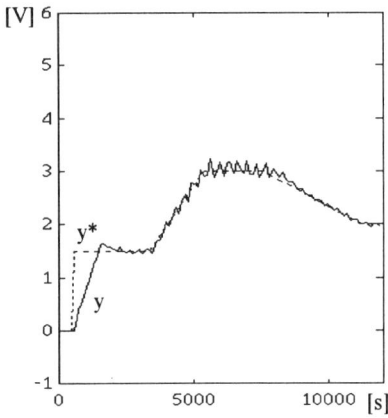
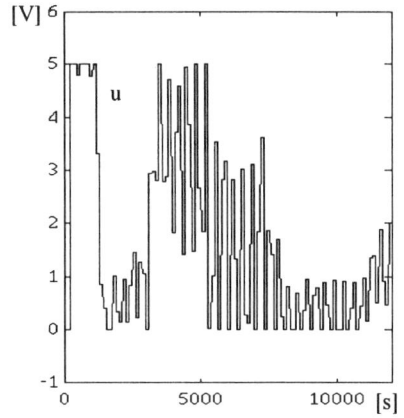

(a)   (b)

Figure 10.5   *Simulation results of the predictive control loop (λ =0.001)*
(*a*) closed-loop response;
(*b*) control variable evolution

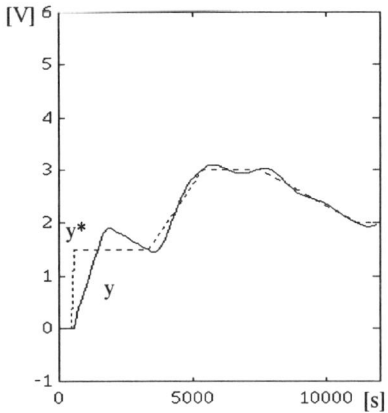
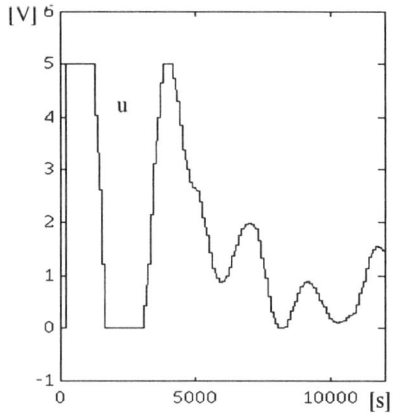

(a)   (b)

Figure 10.6   *Simulation results of the predictive control loop (λ =0.01)*
(*a*) closed-loop response
(*b*) control variable evolution

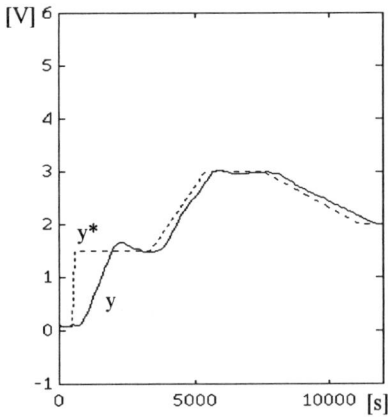 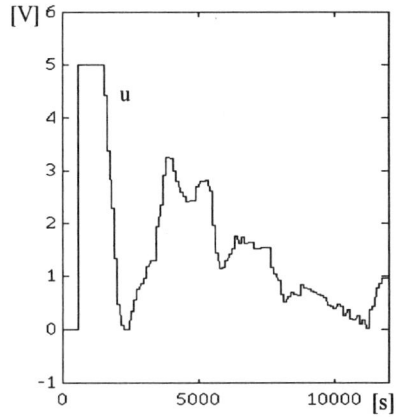

(a)   (b)

Figure 10.7   *Real-time results of the PID control loop*
                (*a*) closed-loop response
                (*b*) control variable evolution

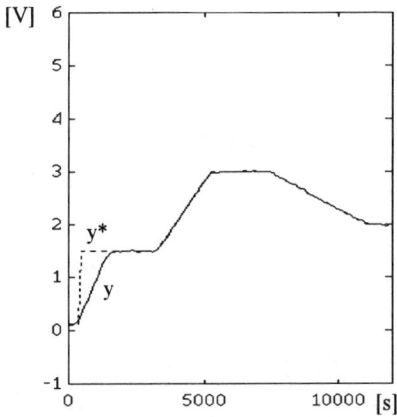 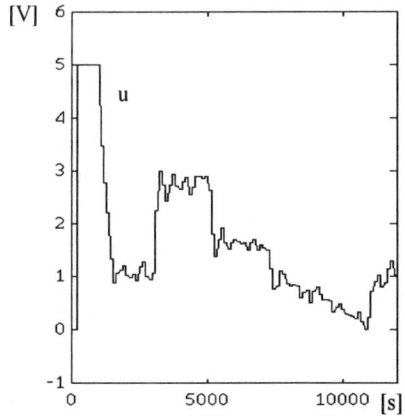

(a)   (b)

Figure 10.8   *Real-time results of the predictive control loop ($\lambda = 0.002$)*
                (*a*) closed-loop response;
                (*b*) control variable evolution

10.8 Conclusions

Temperature control of a thermal treatment furnace with PID and predictive algorithms, respectively, has been presented. Experimental results show a better behaviour from predictive control because of its ability to deal with time delay processes. It is worth noting that predictive control is strongly sensitive to the value of the weighting factor λ.

10.9 References

1   DJEBARA, K., DAHHOU, B., BABARY, J. P., KHELLAF, A. and GARIBAL, C.: 'Partial state model reference adaptive control of a rapid thermal processor', *Int. J. Adaptive Control and Signal Processing*, 1993, **7**, pp. 45–61

2   GAWTHROP, P. J., NOMOKOS, P. E. and SMITH, L. S. P. S.: 'Adaptive temperature control of industrial processes: a comparative study', *IEE Proc. D*, 1990, **137**, (3), pp. 137–144

3   ZIEGLER, J. G. and NICHOLS, N. B.: 'Optimum settings for automatic controllers', *Trans. ASME*, 1943, pp. 759–768

4   KAYA, A. and SCHEIB, T. J.: 'Tuning of PID controls of different structures', *Control Engineering*, July 1988, pp. 62–65

5   ZHUANG, M. and ATHERTON, D. P.: 'Automatic tuning of optimum PID controllers', *IEE Proc. D*, 1993, **140**, (3), pp. 216–224

6   TAKAHASHI, Y., CHAN, C. S. and AUSLANDER, D. M.: 'Parameter-einstellung bei linearen DDC-Algorithmen', *Regelungstech. Prozessdatenverarb.*, 1971, **19**, pp. 237–244

7   GOODWIN, G. C. and SIN, K. S.: 'Adaptive filtering prediction and control', (Prentice Hall, 1984)

8   LAZAR, C., PASTRAVANU, O. and VOICU, M.: 'Laboratory equipment and process set-ups for control systems', *European Seminar on Automatic and Control Technology Education*, Dresden, 1993

Chapter 11

State-space adaptive control for nonlinear systems

K. Janiszowski and M.Olszewski

11.1 Introduction

The integration of a microprocessor and a fine measurement system with a pneumatic cylinder (controlled by a proportional valve) creates a new, powerful servomechanism with adaptive features which can be applied in positioning units of manipulators or robots. The application of such a system introduces certain control problems, which can be solved only by means of modern control, e.g. identification of a model of the plant, estimation and compensation of nonlinearities, design of a state-space controller, state-space reconstruction and, finally, adaptive control of the system. Therefore a computer control system for positioning, using a pneumatic cylinder, is a useful example for presentation of modern control problems.

A pneumatic cylinder is a system consisting of a mass suspended on two air springs and subjected to a friction. It is a well-known mechanic oscillating system with an input in the form of a force of air pressure on the piston surfaces. Movements in a linear system like this are best controlled by a state-space controller. Problems begin at state reconstruction: usually only the piston's position is measured; velocity and acceleration have to be either computed by differentiation of position or estimated by an observer algorithm. The next problem appears with the control valve — usually it has a nonlinear characteristic and exhibits hysteresis. The cylinder is not uniquely defined either; variable mass, cylinder volumes, air temperatures and, most of all, variable friction conditions make its description nonlinear and nonstationary. An analytical analysis yields results far from its real dynamics.

The only method of obtaining a reasonable approximation of this system's dynamics is to make a statistical estimation, but even this method is limited and introduces many problems in practice.

The design of the controller is not completely solved, either. The design of a state-space controller is easy but the limitations of air flows in the control valve make the control nonlinear. The next problem is variation in the plant (e.g. mass of the moving parts) and the need for adequate adaptation rules. The control task is not unique: a position has to be achieved either without overshoot or with a small, acceptable overshoot. A zero steady state error condition can be important or, alternatively, some error is acceptable and response time to the new position may be critical. In the case of real applications, another problem is adaptation to variable working conditions. Where the load is changed, the adaptation has to be very fast — in the same movement. In the case of changed temperature or lubrication conditions, the adaptation can be slow and performed in quite a different way. All these problems can be considered in different applications. Hence a wide range of different algorithms can be applied, investigated and examined with students.

A short discussion of a model of the system, based on description of the phenomena, is given as an introduction to this chapter together with preparation for subsequent identification of a statistical plant model, which will next be applied for the design of the controller.

11.2   Models of piston movements in a pneumatic cylinder

A pneumatic cylinder is a system of two volumes V_1 and V_2, where a difference between air pressures p_1 and p_2 acting on surfaces A_1 and A_2 results in movement of the piston with mounted mass m. The state in both volumes depends on air temperatures of ϑ_1 and ϑ_2, air flows m_1 and m_2, and exponents n_1 and n_2 of polytropic conversion. The dynamics of the system can be described by the piston velocity v and the air pressures in both volumes [1].

$$\frac{d}{dt}v(t) = [-F(v(t)) + A_1 p_1(t) - A_2 p_2(t)]/m$$

$$\frac{d}{dt}p_1(t) = -\frac{n_1 p_1}{V_1}\frac{d}{dt}V_1(t) + \frac{n_1 R\theta}{V_1}\frac{d}{dt}m_1(t) \qquad (11.1)$$

$$\frac{d}{dt}p_2(t) = -\frac{n_2 p_2}{V_2}\frac{d}{dt}V_2(t) + \frac{n_2 R\theta_2}{V_2}\frac{d}{dt}m_2(t)$$

where R denotes the gas constant and $F(v)$ is the velocity-dependent friction of the moving parts. The linearisation of Equation 11.1 for a given position (volumes V_{10}, V_{20}) and air state (pressures p_{i0} and temperatures ϑ_{i0}), with substitution of the pressure difference $p_1 - p_2$ by the piston acceleration a, leads to the following equations:

$$\frac{d}{dt}\begin{bmatrix} x \\ v \\ a \end{bmatrix} = \begin{bmatrix} 0, & 1, & 0 \\ 0, & 0, & 1 \\ 0, & -\omega_o^2, & -2D\omega_o \end{bmatrix}\begin{bmatrix} x \\ v \\ a \end{bmatrix} + \begin{bmatrix} 0 \\ 0 \\ C\omega_o^2 \end{bmatrix}u \qquad (11.2)$$

$$y = \begin{bmatrix} 1, & 0, & 0 \end{bmatrix}\begin{bmatrix} x, & v, & a \end{bmatrix}'$$

where x is the piston position and u denotes control of the proportional servo valve. The valve is considered as an inertia-less element; its cut-off frequency is approximately 10 times greater than ω_o.

In transformation of Equation 11.1 into 11.2, the friction F is substituted by a linear function of velocity v. The system in Equation 11.2 is now described by the following parameters gain C, eigenfrequency ω_o and damping D, which are dependent on the parameters of Equation 11.1 as follows:

$$C\omega_o^2 = \frac{RA\theta_0}{m}\left[\frac{n_1 k_M}{V_{10}} + \frac{n_2 k_M}{V_{20}}\right] \qquad 2D\omega_o = \frac{k_F}{m}, \qquad (11.3)$$

$$\omega_o^2 = \frac{A^2}{m}\left[\frac{n_1 P_{10}}{V_{10}} + \frac{n_2 P_{20}}{V_{20}}\right]$$

for surfaces $A_1 = A_2 = A$. The factor k_F expresses the influence of friction F and k_M is the coefficient of flow characteristic of the valve. The parameters (C, ω_o, D) are dependent on the piston's position (V_{10}, V_{20}) and the mass m and will vary with the piston's movements.

The transfer function for the system in Equation 11.2 is equal to

$$G_{vu}(s) = \frac{C\omega_o^2}{s^2 + 2D\omega_o s + \omega_o^2} \qquad (11.4)$$

The values of system parameters (C, ω_o, D) depend on construction and are usually within the ranges $C \in <0.15, 1.5 \text{ m/s/V}>$ (at ± 10 V valve input), $\omega_o \in <10\,\text{rad/s}, 60\,\text{rad/s}>$ and $D \in <0.1, 1.5>$.

In the case of computer control a discrete-time model has to be introduced. It has been observed that the sampling time T for efficient control has to be within the interval <1 ms, 5 ms$>$. These values are small in comparison with those advised in the literature, e.g. Reference 2, but the total control period is approximately 200–300 ms and the piston brakes in approximately 50–80 ms (see Figure 11.2b), hence greater values of T will introduce poorer control results.

The discrete-time representation in Equation 11.4 can be considered as a form of difference equation

$$v(k) = a_1 v(k-1) + a_2 v(k-2) + b[u(k-1-d) + u(k-2-d)] \qquad (11.5)$$

where k is a discrete-time value, $t = kT$, and d is a time delay $\tau = dT$. The coefficients a_1, a_2 and b are determined by iterative estimation. For sampling time $T=0.33$ ms and the values of frequency ω_o mentioned, these coefficients will have practically the same values [3].

The system parameters (C, ω_o, D) can be derived from the estimations as follows:

$$C = \frac{2b}{1-a_1-a_2} \qquad D = (1+a_2)/[(1-a_1-a_2)((1+a_1-a_2)]^{1/2} \qquad (11.6)$$

$$\omega = 0.5T[(1+a_1-a_2)/(1-a_1-a_2)]^{1/2}$$

This description (Equations 11.3–11.6) is an approximation valid for a piston velocity high enough to exclude an adhesive force of the piston to the cylinder walls that occurs at low velocities. A more real function of the friction force F is presented in Figure 11.1, where F_t is the friction of the seals and f_c is the unknown coefficient of Coulomb friction.

The rapid increase of the friction at low velocities results in a stick-slip effect in control. Other friction effects, such as different tightening forces of the seals in cylinder volumes and frequent asymmetric construction, introduce hysteresis in the friction characteristic and asymmetry in gain.

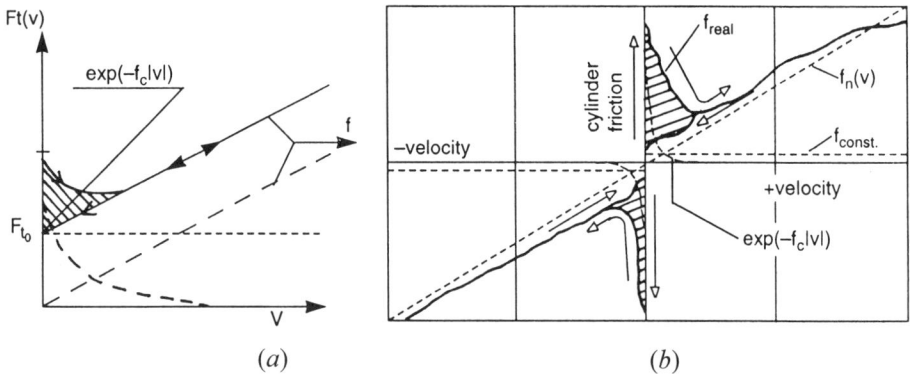

(a)   (b)

Figure 11.1   *Friction force against piston velocity*
(a) expected
(b) measured by experiment

The influence of these factors can be observed in Figure 11.2, which shows measured pulse results of the piston velocity at different valve controls. A distortion of dynamic behaviour at low velocity, Figure 11.2*a*, where the piston was stopped for a while, is yielded by the adhesive friction and is so strong that the piston could not drive smoothly through the cylinder.

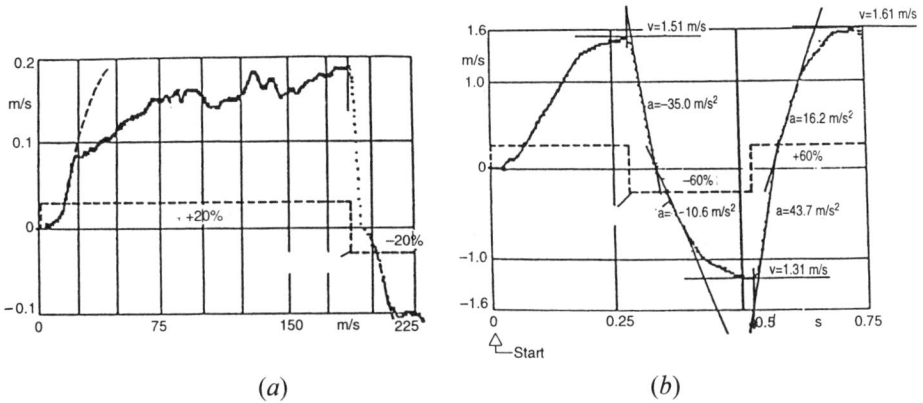

(*a*)   (*b*)

Figure 11.2   *Pulse response of system*
    (*a*) *for 20% changes in control*
    (*b*) *for 60% changes in control*

Figure 11.2*b* shows the asymmetry of fast piston movements. Both figures exhibit a large gap between the transients expected from Equation 11.4 and the observed results. There is a visible difference in gains (ratio of velocity to input signal) in both cases, introduced by nonlinear control of the servo valve. All these effects cause the piston's features to be far from those derived from Equations 11.2 and 11.3, but the main character of system remains — it is a mass suspended on two air springs and moving in the presence of friction. The representation (equation 11.4) is useful but only approximate and has to be adapted to actual conditions.

11.3   Adaptive control system for a pneumatic cylinder

The controllers in classic servomechanisms either have additional velocity feedback or, as in hydraulic systems, a feedback in the form of all three states: position, velocity and acceleration. This feedback yields good dynamics for a closed-loop system with a controller algorithm designed for pole placement.

A full scheme for state-space adaptive control for a nonlinear positioning system is presented in Figure 11.3. A compensator for valve nonlinearity is introduced for correction of the valve characteristics.

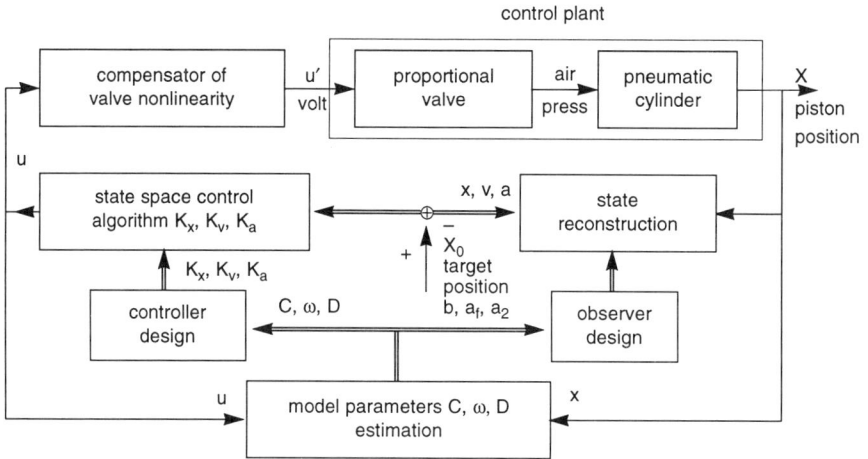

Figure 11.3   Adaptive control system for positioning unit with pneumatic drive

11.3.1 State reconstruction

The state evaluation can be arranged either by direct differentiation of position or by state observers. The first method, based on a simple scheme,

$$v(t) = v(kT) = \frac{x(kT) - x(kT - T)}{T} \tag{11.7a}$$

$$a(t) = a(kT) = \frac{v(kT) - v(kT - T)}{T} \tag{11.7b}$$

introduces high errors at small values of sampling time T. These errors are due to amplification of resolution errors by a factor of $1/T$. For poor measurement systems, errors in the acceleration signal can be higher than the signal itself (errors in estimation of acceleration are magnified by factor of $1/T^2$).

A simple way to decrease these errors is to introduce a differentiation interval $T_{diff} \gg T$ for difference operation and then

$$v(t) = v(kT) \cong \frac{x(kT) - v(kT - T_{diff})}{T_{diff}} \qquad (11.8a)$$

$$a(t) = a(kT) \cong \frac{v(kT) - v(kT - T_{diff})}{T_{diff}} \qquad (11.8b)$$

When T_{diff} equals nT, the quantisation error in signal $a(kT)$ is decreased by a factor of n^2. This approach is simple but yields a time delay approximately equal to $0.5\,T_{diff}$. For very fast systems ($\omega_o > 40$ rad/s), the differentiation interval T_{diff} has to be less than 4 ms. For slower systems satisfactory control results can provide $T_{diff} = 4$–8 ms.

The differentiation scheme of Equation 11.8 gives good results in cases where the resolution of the measurement system is better then 10 μm. In the case of poorer resolution, observer schemes, as discussed in Chapters 5 and 7 and in Reference 4, can be applied. The best results were observed by the application of Luenberger observers or a mixed predictor-derivation scheme, where the velocity $v(k)$ was determined from Equation 11.8, and for estimation of the acceleration the state equation (11.2) or its discrete-time form is used:

$$a(k) = -2\alpha\beta v(k-1) + [1 - \alpha T - 2\beta(1 - \beta)]a(k-1) + 2\alpha\beta Cu(k-1) \qquad (11.9)$$

with

$$\alpha = 0.5\omega_o T \qquad and \qquad \beta = 1 - D\omega_o T.$$

11.3.2 State space control algorithm

The most important task in design of positioning systems is so-called point to point control (P-P control): the servomechanism has to steer a moving part to a given place as fast as possible, without control error and with negligible overshoot. Transients of P-P control are of secondary importance.

There are different approaches to the design of state-space P-P control feedback, e.g. References 2 and 4. For the case investigated, the most useful are pole-placement algorithms. They assure very good transient responses with negligible overshoot or undershoot and low control errors. This is confirmed by the considerable experience from hydraulic and pneumatic control systems, e.g. Reference 5. The controller action is determined by the relationship

$$u(k) = -K_x(w - x(k)) - K_v v(k) - K_a a(k) - offset \qquad (11.10)$$

where w denotes a target position for the piston, K_x, K_v, K_a are controller gains and *offset* is a value introduced for compensation of asymmetry of the valve.

The gains (K_x, K_v, K_a) are determined from the system parameters (C, ω_o, D) and the predefined poles s_1, s_2, s_3 of the closed-loop system. The usual approach is to set one real pole and a pair of conjugate poles with moderate damping. The poles have to be dependent on the eigenfrequency ω_o and damping factor D of the system (Equation 11.4)

$$s_1 = \operatorname{Re} s_2 = \operatorname{Re} s_3 = -1.5\omega_o, \qquad \operatorname{Im} s_2 = -\operatorname{Im} s_3 = -0.75\omega_o. \qquad (11.11)$$

A requirement for smaller control errors yields other choice of poles [5]

$$s_1 = -\kappa_1 \omega_o (1 + D^2)^{1/2}$$
$$\operatorname{Re} s_2 = -\kappa_2 \omega_o (1 + D^2)^{1/2} \qquad \operatorname{Im} s_2 = -\kappa_3 \operatorname{Re} s_2. \qquad (11.12)$$

The coefficients $\kappa_1, \kappa_2, \kappa_3$, which depend on the system parameters (C, ω_o, D), optimised for low sensitivity in respect of damping D, are as follows:

$$\kappa_1 = \lambda / \omega_o - 1 + D, \qquad \kappa_2 = 1.3, \qquad \kappa_3 = 10 / (3\omega_o + 16D) \qquad (11.13)$$

where λ is a coefficient which can vary within the range 50–1000. For lower values of λ, valve action is slow but the control error is high. High values of λ involve oscillatory action of the valve but decrease the error. The controller gains are as follows:

$$K_x = -s_1 (\operatorname{Re} s_2 + \operatorname{Im} s_2) / C\omega^2$$
$$K_v = (2s_1 \operatorname{Re} s_2 + \operatorname{Re} s_2^2 + \operatorname{Im} s_2^2 - \omega^2) / C\omega^2 \qquad (11.14)$$
$$K_a = ((\kappa_1 + 2\kappa_2)\omega (1 + D^2)^{1/2} - 2D\omega) / C\omega^2$$

11.3.3 Model parameter estimation

For estimation of the coefficients in Equation 11.5, different recursive schemes can be used but, in the case investigated, there are no visible external distortions of the output signal, hence the simple least-squares recursive method (RLS), as described in chapter 1, can be used without risk of losing estimation accuracy. Figure 11.4 illustrates typical behaviour of on-line estimations. Figure 11.4*a* shows estimations of the system parameters (C, ω_o, D) calculated under standard controller action.

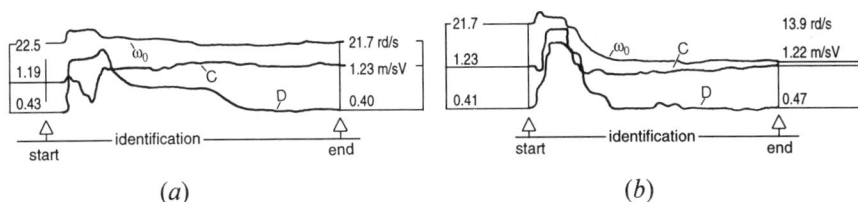

Figure 11.4 On-line estimation of model parameters within one braking interval

System parameters C, ω_o, D
(a) for the valve-cylinder system with a 10 kg mass load
(b) for the same system with change of load from 10 to 30 kg

The on-line calculations are started when the braking action of the piston is observed and are performed until the piston velocity attains 10% of its maximum value. One can observe transients of estimations due to 'refreshing' of the covariance matrix [6]. Figure 11.4b shows estimations of the system parameters in the case of a change in mass load from approximately 10 to 30 kg. The system eigenfrequency ω_o has to decrease, see Equation 11.3.

11.3.4 Compensation of valve nonlinearity

As has been mentioned, the valve characteristics supplied by manufacturers are not very precise and are not measured in conjunction with the cylinder investigated. For small cylinders the 'proportional' valve can act in a highly nonlinear manner, see Figure 11.5.

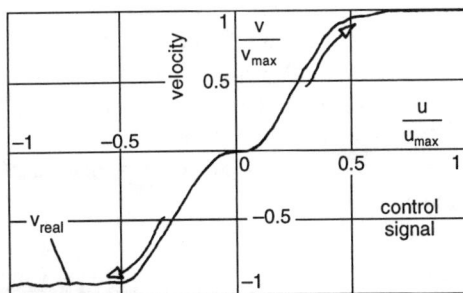

Figure 11.5 Efficiency of valve action for a cylinder of 25 mm diameter (measured in both directions)

This characteristic is for a valve with negative overlap. The first part of this characteristic determines an offset value introduced in Equation 11.10.

Compensation for valve nonlinearity can be introduced in a control system as a compensator for the valve characteristics or as part of the difference model (Equation 11.5) where terms b_i present nonlinear functions of u.

11.4   Fuzzy control system for a pneumatic cylinder

The alternative method of controller design is a fuzzy approach to the controller's task. Since both methods can be applied to the same equipment, it is worth outlining this approach.

Fuzzy design, e.g. [7] and chapter 13, does not require any information about system parameters. The tuning of the controller is based on observation of the following parameters of control action: an overshoot (or undershoot), magnitude of the negative minimum of the piston velocity in the braking phase and a number of extrema in deceleration. These parameters are evaluated for each movement and are then applied for determination of the gains (K_x, K_v, K_a) of the controller with fuzzy rules. This approach yields fast control with small overshoots and moderate control error. Its advantage is a simple algorithm suitable for cheap microcomputers but it is less accurate and its adaptation is slow, needing 15–30 movements to tune the controller.

11.5   Laboratory equipment

The laboratory system for nonlinear adaptive control was set up from a mechanical construction of different cylinders of 200–800 mm in length, a proportional valve, an optical measurement system with 2 μm resolution, TMS320C25 signal processor in the interface system and a PC-AT computer.

The mechanical construction of two different stands is displayed in Figure 11.6. The carriage, mounted on the shaft, can be loaded with different masses and is suspended on two massive round shears (shafts from other cylinders) to decrease possible deflection in the case of the higher mass load.

An important hint when constructing the stand is to separate the suspension of the carriage and cylinder.

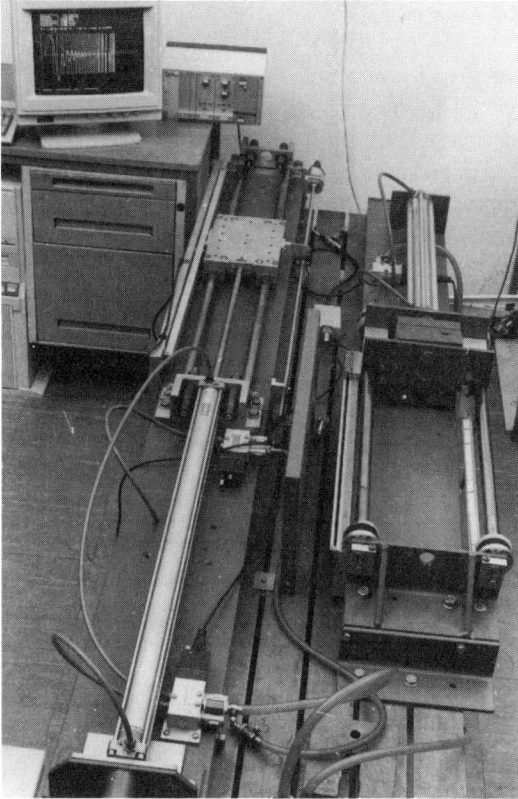

Figure 11.6   Development system for pneumatic positioning systems

The joint suspension (right stand) suffers problems with exchange of cylinders and can yield gripping effects. A much better mechanism is to disconnect the mounting of the cylinder and mounting of the carriage (left stand).

11.6 Experimental scope

The stand described offers the opportunity of different experiments. Most of them have been tried with student or diploma projects and the most interesting have been included in the programme of student laboratory experience.

11.6.1 Determination of static characteristics of the valve-cylinder system

This preliminary research will give a rough estimate of the cylinder gain C and a value of the valve offset. The characteristics should be similar to those shown in Figure 11.5. The experiment must start in the middle of the cylinder and the control of the piston has to be in the form of rectangular pulses for a definite time. After this period (approximately 100 ms) of control equal to value u the corresponding piston velocity is measured and the valve efficiency is determined. The most relevant educational aspect at this point is to show the students that the static nonlinearity of the control valve depends on the cylinder controlled. The differences between these characteristics for small cylinders, e.g. of diameter 20 mm, and the bigger ones, e.g. of 40 mm, are spectacular.

11.6.2 PID control

Tuning of the controller settings can be based on the piston velocity determined in the previous step from the results of the pulse tests, considered as an approximation of a step response. The controller settings can be obtained from different tables, e.g. References 2 and 8, and final tuning can be made iteratively by testing different settings and observation of results. The advised sampling time is 2–10 ms.

A typical observation is a stick-slip effect, which worsens the positioning action and yields the conclusion that this type of control is unsuitable.

It is possible to introduce a PD algorithm into the control loop, but then high control errors will be observed after tuning.

11.6.3 On-line system identification

Identification has to be based solely on the braking phase of the movement. The experiments can first be performed with full control of the valve and repeated 3–5 times for different piston positions, for estimation of average values of the model parameters (C, ω_o, D). The average system parameters derived from Equation 11.6 are then used for determination of the state-space controller constant: Equation 11.13, for $\lambda=100$ in Equation 11.12.

For comparison, the same procedure can be repeated but with limitation of maximum controller output to 60% and only 30% of full control of the valve. The results will confirm the nonlinearity of gain C.

Of further interest are estimations of model parameters in experiments where some parameters of system, e.g. the mass load of the moving carriage, are changed, the estimated parameters have to change in the relevant direction (see Figure 11.4b).

11.6.4 State space control

The control experiments can be arranged in the form of cycles of short, medium and long piston movements.

The movements are very fast, hence it is advisable to build into the control program a package for automatically saving the input and output data for each movement onto disc for subsequent analysis. Plotting these signals for display in real time, as the controller acts, is difficult. The advised sampling time is 1 ms.

The gains of the state-space control algorithm can be derived from Equations 11.11–11.13 with system parameters determined in the previous section. The factor λ in Equation 11.12 should first be set to 50 and then increased to 600; the differences in control error are visible (at $\lambda=600$, control errors are roughly 10% of those at $\lambda=50$).

Some results of the state-space control are depicted in Figure 11.7a, which shows signals from positioning which is practically time-optimal. The position of the piston and its velocity are shown in the upper part. In the lower diagram the controller input, which is nearly time-optimal in sequence, (first full acceleration and next full braking action) and enlarged control error (100 μm amplitude) are shown. The control time (measured up to 100 μm zone) is approximately 168 ms.

Typical positioning results (for the same position) are shown in Figure 11.7b, obtained with a controller designed using Equations 11.12 and 11.13 in the adaptive control of the system. The profiles of position and velocity are very similar to those of Figure 11.7a, but the controller output and control error (lower part) show differences. The controller cannot establish braking as efficiently and the control time is visibly longer — approximately 264 ms. The controller output is visibly more active than in the previous case but it is adaptive for sudden changes of moving system, e.g. variation of load or lubrication conditions.

Positioning results for a strongly overdamped controller are given in Figure 11.7c. This effect can be achieved when the factor κ_3 in Equation 11.13 is decreased to achieve small or zero overshoots. The piston's movement may be braked too early and has to be accelerated (see the velocity signal) in some parts of the movement. Positioning is very slow — it takes approximately 345 ms to reach the correct position. A small overshoot has appeared and the valve activity is too intensive, and can lead to shortening of the time between valve overhauls.

11.6.5 Adaptive state-space control

The adaptation in the system described can be performed in another way as for classic adaptive systems. On-line model estimation is active only in the braking phases of the controller action for a specified piston position. After each start, some transients in the coefficient estimations appear and only after 40–50 calculation steps, (e.g. Figure 11.4a), the model is suitable for the design of the controller. A condition for change in the controller gains is based on velocity value: when the piston velocity in the braking phase of the movement is decreased to 10 or 5% of the maximum value, a revision of control gains can be made.

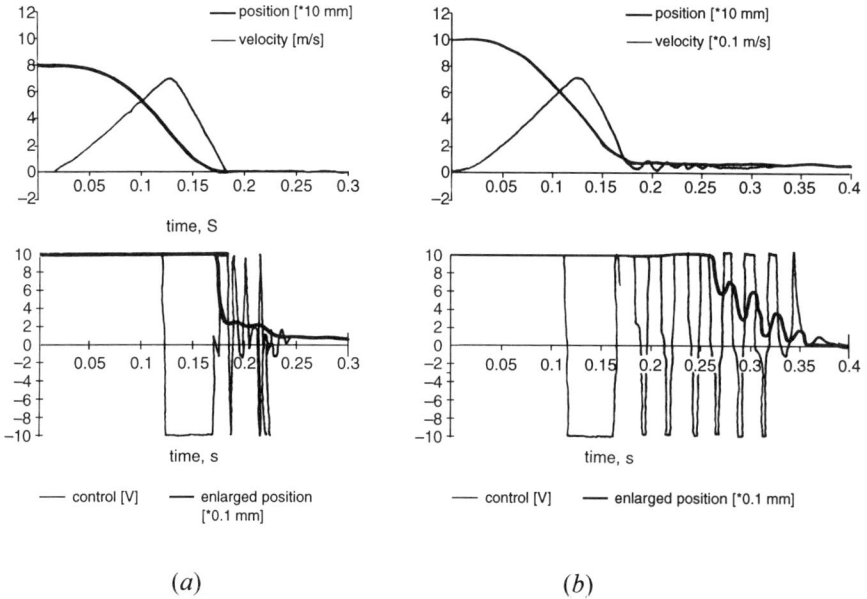

(*a*)   (*b*)

Figure 11. 7   Transients of position and piston velocity (upper part) and control error with controller output (lower part)
(a) For time optimal controller gains
(b) For typical adaptive controller settings

11.6.6 Compensation for nonlinearity

Compensation is introduced in the form of a static function, being the inverse of recorded valve nonlinearity (see Figure 11.5). The compensation for nonlinearity and its positive effect is best observed during positioning with small cylinders and small loads. The nonlinearity of the control valve, in the absence of compensation, can then significantly destroy positioning quality and introduce long, weakly damped oscillations into controller action. In these cases (Figure 11.5), 30% of controller input is practically equivalent to 100% of the control needed to induce oscillations at small inputs of the controller valve, due to, for example, errors in state (acceleration) estimations. For larger cylinders this effect is less visible — movements of the piston are slower and errors in state estimations do not influence controller output so strongly.

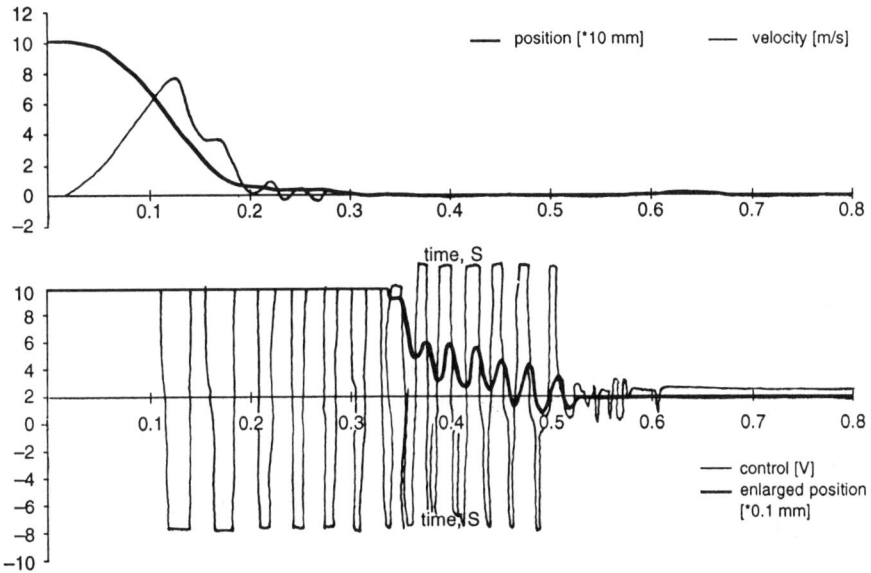

Figure 11.7 c   Transients of position and piston velocity (upper part) and control error with controller output (lower part) for an overdamped controller

The limited influence of valve nonlinearity observed was not expected at first; our expectations had been more pessimistic. The effect can be explained after inspection of Figure 11.7. The controller output, in the most important intervals of control, reaches its limits — hence compensation for the nonlinearity is not required. Its positive effect can only be observed in those intervals when it does not attain its limits, i.e. at the conclusion of movement. Hence, the positive influence of compensation is observed only in more sensitive, small cylinders.

Estimation of the nonlinearity is vital in follow-up control, when the piston control does not attain its limits during the control activity.

11.7   Conclusions

The system described for adaptive control creates a variety of problems, which can be demonstrated on the same apparatus. Many algorithms in modern control have to be applied if satisfactory control results are required. The classic PD or PID control is not useful in the case considered. Only state-space control can fully cope with the problem. Other approaches, such as dead-beat control, can be considered too; however, this type of controller is very sensitive to distortion of the controller output, which is evident in the valve action.

Various schemes for state reconstruction can be used but, even in the case of poor resolution, a simple numerical differentiation is most useful.

Different methods are used for state-space controller design but pole placement with the proposed choice of closed-loop poles has resulted in good control parameters and is advised for this special application.

The introduction of nonlinearity compensation does not introduce much difference in the case of exact control (with a high value of λ). Then the controller gains are very high and the controller output attains its limits throughout almost the complete braking interval. The compensation for nonlinearity is necessary in tracking problems, when the piston velocity is controlled.

The efficiency of the adaptation method described is observed in the controller design. Good knowledge of the system parameters leads to stable control with high controller gains K_x which yields very low control error. If the control error is less important, but, for example, control time should be minimised, other strategies can be developed.

The most important feature of the control system presented is the variety of control problems that are introduced by a simple element such as a pneumatic cylinder, along with its possibilities for systematic training of different algorithms. For this reason we consider that it should become a standard experimental problem, similar to the inverted pendulum or multi-tank level and temperature control.

11.8 Acknowledgements

The authors would like to thank the Polish National Research Committee and Festo Co. for support of the research programme on which this work is based.

11.9 References

1   OLSZEWSKI, M.: 'Konzept der Zustandregelung fur schwachgedampfte Fluidantriebe', *Olhydraulik und Pneumatik*, 1991, **35**, pp. 932–941

2   ASTROM, K. and WITTENMARK, B.: 'Adaptive control', (Addison Wesley, Massachusetts, 1989)

3   JANISZOWSKI, K.: 'A modification of the Tustin approximation', *IEEE Trans.*, 1993, **AC-38**, pp. 1313–1317

4   ACKERMANN, J.: 'Sampled data control systems' (Springer, New York, 1985)

5   OLSZEWSKI, M. and JANISZOWSKI, K.: 'A survey on pneumatic positioning systems', *National Conf. on Control*, Gdansk, 1994 (in Polish)

6   ISERMANN, R.: 'Identifikation dynamischer Systeme', (Springer Verlag, Berlin, 1988)

7   KLEIN, A.: 'Einsatz der Fuzzy-Logik ...', PhD thesis, 1993, Aachen, Germany

8   TAKAHASHI, Y., *et al.*: 'Control and dynamic systems' (Addison Wesley, Massachusetts, 1970)

Chapter 12

Distributed process control

B. Rohál-Ilkiv, P. Zelinka and R. Richter

12.1 Introduction

Distributed parameter processes are dynamic systems the state of which depends
not only on time but also on spatial coordinates. These processes are frequently
encountered in important engineering problems. This chapter presents a practical
approach to adaptive control of spatial temperature distributions in real thermal
systems. Spline functions are used to move from the infinite-dimensional primary
process representation to the finite-dimensional (lumped parameter) process
description. In this way approximate, real-time models of the distributed processes
can be constructed and identified using standard recursive methods. Application to
distributed and boundary control is outlined. The resulting control algorithms can
be verified and tested by various experiments with two simple laboratory-scale
models of thermal distributed processes. The underlying principles of the models
are explained here and their use experimentally is shown.

12.2 Motivation

Heating various solid materials is a typical technological operation in industry. In
many of these operations the aim is to remove the material once its core
temperature has reached a specified value, or once the temperature within the
material has reached a particular spatial distribution. Moreover, the heating
process should be as fast as possible, efficient to save energy and optimal from
both technological and ecological points of view. Modelling of these processes
naturally leads to problems of distributed parameter systems. The states of such
systems are always a function of spatial positions, that is to say they are
distributed. Frequently, only some of these distributed state variables are fully
accessible for direct measurement. In this case the point is how to control the
unmeasured distributed state variables utilising only output data measured on the

distributed system and the knowledge of the physical laws governing the process at hand.

The thermal state within an ingot during the heating operation in a multizone reheating furnace is considered (for a simple scheme of the furnace, see Figure 12.1). The main aim is to reach a specified internal temperature in the ingot at the furnace exit which is necessary for hot working. The main difficulty is that the internal temperature cannot be obtained by routine measurement; it can only be predicted using a suitable mathematical model.

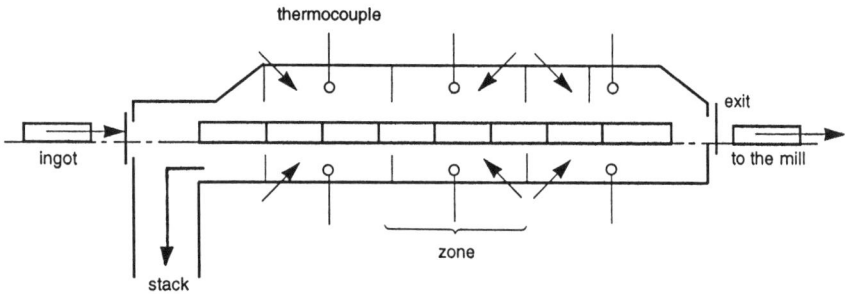

Figure 12.1   Multizone reheating furnace of the pusher type

Further problems arise in temperature control of the furnaces from:

• nonlinearity of the controlled system and its asymmetrical behaviour, due to different time constants in heat supply and heat removal;
• variations in process dynamics due to irregular movement of the ingots through the furnace;
• time delays due to the distributed parameter nature of the process.

The goal of this chapter is to present basic ideas for an adaptive control scheme which can handle the problem of control of spatial temperature distributions in real thermal systems. For the case of boundary control, the principle of the control scheme consists of varying the boundary conditions (surface temperature T_s of heated material, see Figure 12.2) that govern the pattern of heat transfer into the heated material until it is calculated that a required value of internal temperature T_c has been reached. Control is achieved not by feedback but by maintaining a pre-calculated (reference) temperature–time profile at the boundary of the heated object. The feedback is then applied to control of the furnace temperature T_f or the surface temperature T_s if this temperature is accessible to direct measurement.

The theory of the distributed parameter processes is rather complicated and involves many abstract notions and complex methods. Any practical illustration of the theoretical principles in the laboratory would seem to be very useful to obtain a better physical insight into the problems. In this chapter two laboratory equipments supporting topics of distributed parameter processes are described.

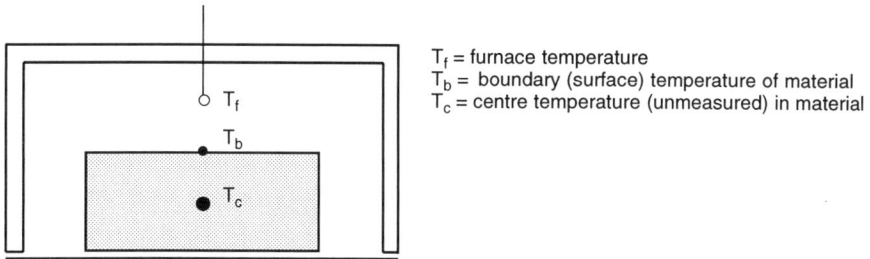

T_f = furnace temperature
T_b = boundary (surface) temperature of material
T_c = centre temperature (unmeasured) in material

Figure 12.2   Heating a solid material in a furnace

12. 3 Technical approaches to distributed process control

The dynamic behaviour of distributed parameter processes is described by partial differential, integral or integro-differential equations and, in many cases, by general functional equations. These equations are derived from certain fundamental laws and hypotheses based on physical arguments. The distributed process is defined on a certain spatial domain which can be fixed or variable with time. A control variable which is acting over the whole spatial domain is called distributed control and a control variable acting on the boundary of the domain is called boundary control. There are numerous theoretical and technical publications which deal with modelling and control of distributed parameter processes and which describe applications in mechanical, chemical, thermal systems etc., see References 1–5.

Depending on the shape of the spatial domain and the type of equations, the problems encountered in modelling and control of distributed parameter processes are often very complex. Therefore, for engineers, a common way to treat these problems is to use simpler, approximative methods which are often very effective in the process of moving from the infinite-dimensional (distributed parameter) nature of the processes to the finite-dimensional (lumped parameter) process description.

A straightforward technique is to divide the spatial domain into a finite number of subdomains (zones) in which the spatial distribution of variables is neglected. In many cases this so-called zonation technique produces satisfactory results provided the number of zones is not large. A variety of other methods exist to reduce the infinite-dimensional, primary process model to a finite-dimensional one, such as:

- finite difference method
- finite element method
- modal approximation
- Fourier transform method

Mostly these methods mean approximation of the partial differential equations by a set of ordinary differential or algebraic equations. Our obvious conclusion is to treat the distributed parameter problem by substituting such an approximation for it at some stage in the analysis. This might be done at the beginning of the analysis — early approximation — or in the final phases of the analysis — late approximation. Late approximation enables us to solve the problem in an infinite-dimensional framework as far as it is possible. The form of the approximation is fundamental in the real-time implementation and determination of the control law.

12.4 Discussion

The techniques above are classical procedures used by many specialists to compute approximate solutions of partial differential equations (PDEs) or to obtain analytical formulas which can represent the solutions. The starting point of these techniques is a known primary model, mostly in the form of PDEs with specified initial and boundary conditions, which describes the process dynamics. In many practical situations it is difficult to obtain such equations which can represent satisfactorily the controlled process dynamics, especially if we want to take into account:

- time variations in process dynamics;
- process uncertainties; and
- stochastic disturbances.

A simple technique of approximation of the distributed processes using a spline function, suggested in Reference 6, appears to be useful to form simple real-time models. The technique utilises some of the ideas of probabilistic modelling of real systems introduced in Reference 7 and does not rely on PDEs. The resulting finite-dimensional models (of regression type) are especially suitable for adaptive control design. In the following two sections this technique will be described briefly when applied to distributed thermal systems.

12.4.1 Distributed control

Let us consider that the temperature distribution is accessible to direct measurement at a suitable number of points. The essence of the spline technique is simple: at discrete time instants t ($t = 0,1,2,...$) the spatial temperature distribution $y(x,t)$ in the thermal system is approximated via suitable spline functions chosen and the dynamics of the spline coefficients then investigated using measured input/output data. The number of the system outputs fully depends on the desired level of approximation errors.

Let us suppose that the thermal system is controlled by du heating elements acting on a distributed domain which are manipulated by inputs $u_j(t), j = 1,2,...,du.$ In the following text we shall refer to the du dimensional vector $u(t)$

$$u(t) = \left[u_1(t), u_2(t), \ldots, u_{du}(t)\right]^T \tag{12.1}$$

as the input vector of the controlled thermal system. Let us consider that the spatial temperature distribution $y(x,t)$ is measured at dy discrete points with coordinates x_i, $x_i \in [0, L]$, (L is the active length of the system), $i = 1,2,...,dy$, and let us approximate the spatial temperature distribution by the spline function $ys(x,t)$:

$$ys(x,t) = cy^T(t)m(x,t) \tag{12.2}$$

where $cy(t)$ is the dcy-dimensional vector of spline coefficients and $m(x,t)$ is the vector of suitable B-splines chosen. Then the model of the controlled system can be written in the following finite-dimensional form:

$$cy(t) = \sum_{i=1}^{ny} A_i cy(t-i) + \sum_{i=0}^{nu} B_i u(t-i-d) + c + \varepsilon(t) \tag{12.3}$$

where
  A_i, B_i – matrices of unknown model parameters
  $\varepsilon(t)$ – vector of unmeasurable disturbances
  c – absolute term
  ny, nu – model orders
  d – time delay.

It is possible to estimate the unknown parameters of the model (Equation 12.3) by standard recursive methods from the measured input/output data while the distributed nature of the controlled process is kept to a reasonable level.

To determine the optimal values of the control input signals $u_j(t)$ we use the following integral control criterion in which the spatial distribution of the controlled variable is included:

$$J^H = E\left\{\frac{1}{H}\sum_{k=1}^{H}\left[\int_0^L \left[(y(x,k)-w(x,k))g(x)\right]^2 dx + (u(k)-u_r(k))^T Q_u(u(k)-u_r(k))\right]\right\}$$

(12.4)

and:

$$u(k) \in U(k) \qquad k = 1,2,\ldots,H$$

(12.5)

where:

$w(x,k)$	is a reference value for $y(x,k)$
$u_r(k)$	is a vector of reference values for the input $u(k)$
$g(x)$	is the weighting function of the distributed control error
Q_u	is the positive semidefinite input penalty matrix
H	is the control horizon
$U(k)$	is the set of admissible inputs.

If approximations of the spatially distributed functions $y(x,k)$, $w(x,k)$ in the direction x are performed with the help of the spline functions, then criterion 12.4 can be transformed to the finite-dimensional form:

$$J_a^H = E\ \left\{\frac{1}{H}\sum_{k=1}^{H}\left[(cy(k)-cy_r(k))^T Q_y(cy(k)-cy_r(k)) + \right.\right.$$

(12.6)

$$\left.\left.(u(k)-u_r(k))^T Q_u(u(k)-u_r(k))\right]\right\}$$

with condition (12.5) where vector $cy_r(k)$ holds reference values of the spline coefficients, Q_y is a positive semidefinite non diagonal weighting matrix, its elements are defined as integrals of all combinations of the mutual products of any two B-spline functions $B_i(x)B_j(x)$:

$$Q_{y_{i,j}} = \int_0^L B_i(x)B_j(x)g^2(x)dx$$

where z is the dimension of the spline base and $i, j = 1,2,\ldots,z$.

The minimisation of the criterion (12.6) with condition (12.5) can be achieved via the quadratic programming method (a full description is given in References 6 and 8). The resulting multivariable adaptive control algorithm has been applied to control of spatial temperature distribution in a laboratory aerothermal system and implemented to control temperature distribution in an industrial multizone reheating furnace.

12.4.2 Boundary control

The problem of boundary control obviously arises in thermal systems if the spatial temperature distribution is not accessible to direct measurement. Then the controlled thermal system can be decomposed into two subsystems – a subsystem which is easy to control by feedback using a measurable process output and a subsystem with a distributed thermal state — usually a spatial temperature distribution inside a heated material — which is inaccessible to direct measurement. The dynamics of the thermal state can theoretically be described by a suitable distributed parameter model with a boundary excitation performed via the measurable system output. The control task then consists of optimally varying the measurable system output that governs the boundary excitation of the distributed parameter subsystem, until it is calculated that the required shape of the distributed state has been reached. Optimal reference values for the system output, which should be tracked by a controller, are generated using a specific technique for inversion of the distributed parameter models introduced in Reference 9. In the paper cited, the inverse problem is converted to a regularisation problem and is solved by a stepwise technique. This technique seems to be suitable for the on-line control of thermal systems under conditions of stochastic disturbances acting on the controlled systems.

It is useful to combine this technique of inversion with the predictive adaptive controller for feedback control of the measurable system output. The overall control scheme is shown in Figure 12.3. The mechanism of the proposed method can easily be tested using a simple laboratory thermal model consisting of a thin copper bar and heating apparatus suitable for boundary heating of the bar. The spatial temperature distribution in the direction of the bar length is measured using several equidistantly located thermocouples, see Figure 12.4. In the proposed boundary control scheme, the heating apparatus is considered to be the controlled subsystem of the thermal system with input signal $u(t)$. The measurable system output $y(t)$ is the boundary temperature of the bar. This temperature is the manipulated input to the second subsystem — the heated copper bar — where the unmeasured spatial temperature distribution of the bar $s(x,t)$ is modelled by the known equation of heat conduction:

$$\frac{\partial}{\partial t}s(x,t) - a^2 \frac{\partial^2}{\partial x^2}s(x,t) + bs(x,t) = 0$$

$$s(x,t_o) = s_o(x), \quad s(0,t) = y(t), \quad \frac{\partial}{\partial x} s(L,t) = 0$$

$$0 \le x \le L, \quad t \ge t_o, \quad a \ne 0 \tag{12.7}$$

$$a^2 = \frac{\lambda}{c \cdot \rho}, \quad b = \frac{h}{c \cdot \rho}$$

where L is the length of the bar, λ is the thermal conductivity coefficient, c is the specific heat, ρ is the specific mass of the bar and h is the heat-transfer coefficient.

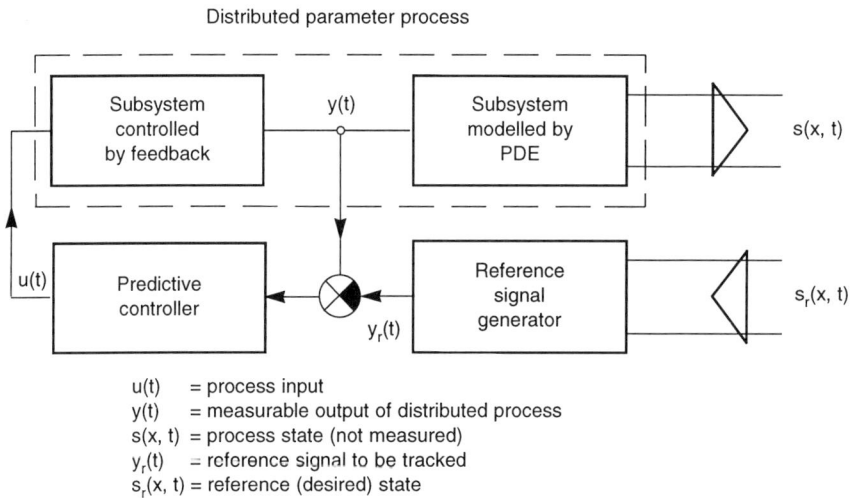

Figure 12.3   Control scheme

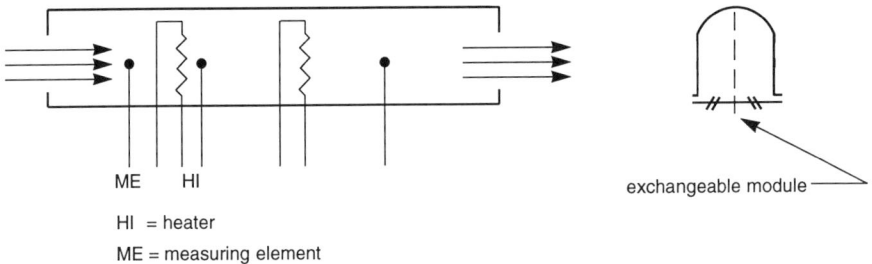

Figure 12.4   Draft of aerothermal laboratory model with cross-section

12. 5 Laboratory set-up

The problems of distributed processes control can be analysed easily and understood better using the following simple laboratory arrangement.

12.5.1 For distributed control

The basic equipment for this experiment consists of:

- aerothermal laboratory model;

- commercial data acquisition card;

- personal computer.

The physical basis of the aerothermal laboratory model is the flow of air through a long tube with a small constant cross-section of semicircular form. The flowing air is heated by several electrical heaters which are distributed along the length of the tube, see Figure 12.5. The active length L of the apparatus in our control laboratory is $L = 600$ mm and the area of the cross-section is approximately 7 cm^2. The temperature of the hot air is measured by bead thermistors. The air flow in the direction of the tube length can be influenced by dividing walls which enable us to divide the apparatus into several zones. The process disturbances are introduced by several inlets of cooling air to the tube. The heating elements, measuring elements, dividing walls and the inlets of cooling air are constructed as simple exchangeable modules which can be placed anywhere in the direction of the apparatus length. Applying these modules we can easily assemble various forms of distributed parameter process.

The system inputs are:

- heating inputs (electrical heaters with maximum power 25 W, the manipulated variable is voltage $u(t) \in [0 \text{ V} \div 10 \text{ V}]$);

- cooling-disturbing inputs (electrical fans).

The system output is a spatial temperature distribution $y(x,t)$, $x \in [0, L]$), $t \geq 0$, in the tube, which is measured by thermistors. This temperature can be represented by:

- one or more variables of lumped character, with values equal to the temperatures measured at given points;

- a suitable function (spline) which approximates the spatial distribution of the air temperature $y(x,t)$ measured in a finite number of points (distributed parameters approach).

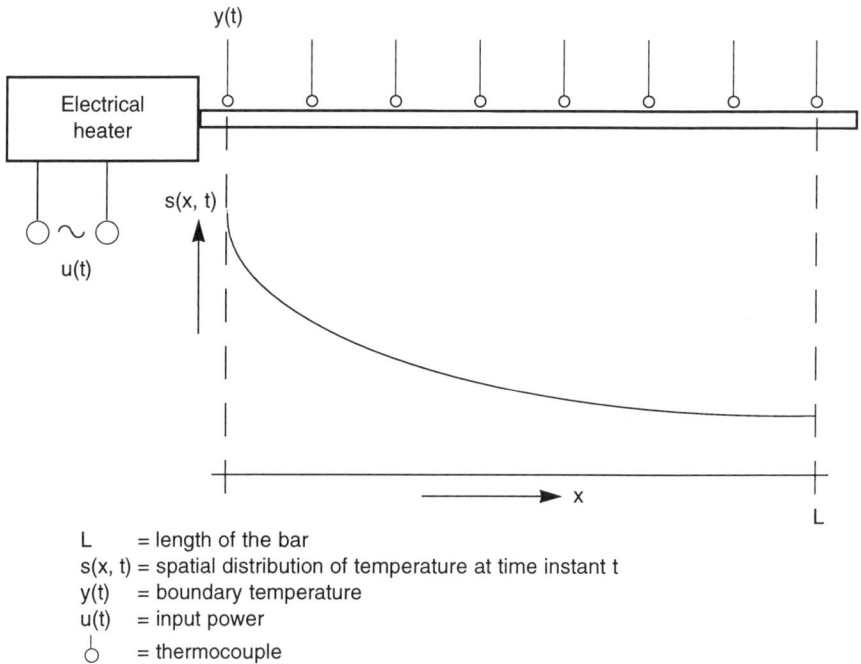

L   = length of the bar
s(x, t) = spatial distribution of temperature at time instant t
y(t)   = boundary temperature
u(t)   = input power
⎤   = thermocouple
⎦

Figure 12.5   Electrically heated thin copper bar

12.5.2 For boundary control

The basic equipment required is:

- electrical heater and copper bar with thermocouples;

- commercial data acquisition card;

- personal computer.

The problem of how to manipulate the unmeasured distributed temperature state $s(x,t)$ can be simulated by simple laboratory apparatus — such as the boundary heated copper bar (see Figure 12.5), which demonstrates a one-dimensional heating problem. For experiments, a copper bar with diameter 5 mm and length $L = 280$ mm is considered. The bar is equipped with eight equidistantly spaced thermocouples.

12.6   Suggested experiments

The following experiments and tasks can be performed with the help of the above laboratory set-up.

12.6.1   Aerothermal process

- Real-time and finite-dimensional modelling of distributed parameter processes based on spline approximation of the spatial temperature distribution;

- Determination of the structure and parameters of the regression models based on measured input/output data;

- Identification of time-varying distributed parameter processes employing various methods of parameter tracking;

- Self-tuning and adaptive control of spatial temperature distributions.

12.6.2   Boundary heated thin copper bar

- Parameter estimation for a one-dimensional heat equation using experimental data;

- Inversion of the heat equation by a stepwise method;

- Design of an optimal boundary control model using the distributed parameter process;

- Design of a controller for the system with considerable time delay.

12.7 Illustrative results

To illustrate the mechanism for the proposed distributed control, consider the following experiment with the aerothermal process. The experiment considers the comparison of distributed control with conventional multivariable LQG control. Let us measure the spatial temperature distribution in the laboratory apparatus using eight equidistantly spaced thermistors. The flowing air is heated by four electrical heaters. Figures 12.6 and 12.7 represent results obtained for multivariable LQG control of the temperature distribution. As the controlled outputs we chose four temperatures measured at points x_2, x_4, x_6 and x_8. The output and reference signals are shown in Figure 12.6, the control input signals in Figure 12.7 and the disturbance signal — air flow — used during the experiment in Figure 12.8. Figures 12.9 and 12.10 give results obtained for control using spline approximation of the measured spatial temperature distribution. The controlled output is now the vector $cy(t)$ of spline coefficients with $dcy=5$. It is easy to see that the spline approximation reduces the effect of noise acting on the temperatures measured and gives smoother control signals.

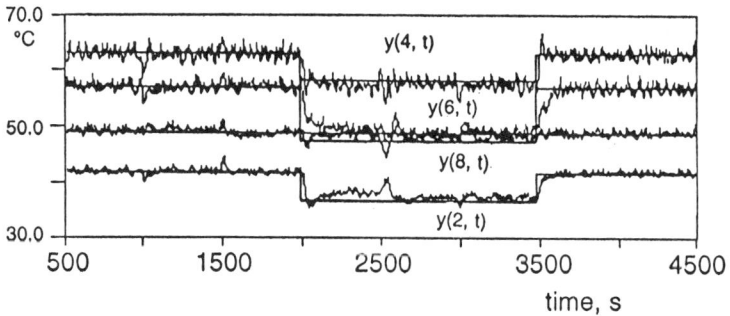

Figure 12.6   Output and reference signals

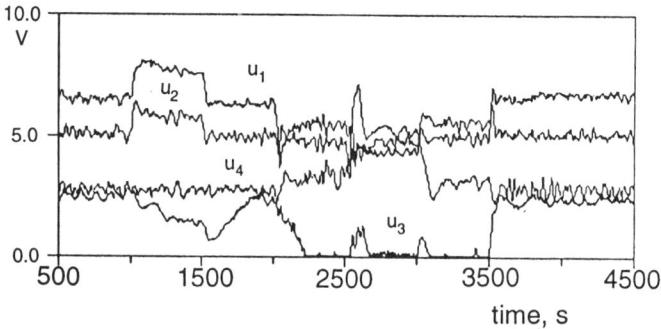

Figure 12.7   Control signals

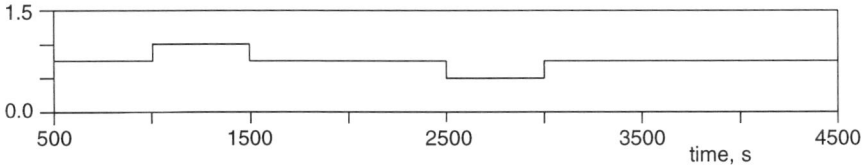

Figure 12.8   Air flow

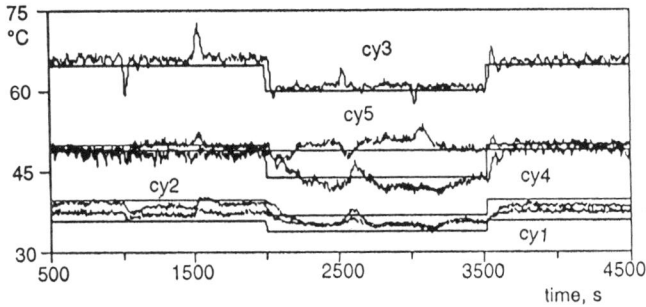

Figure 12.9   Output and reference signals

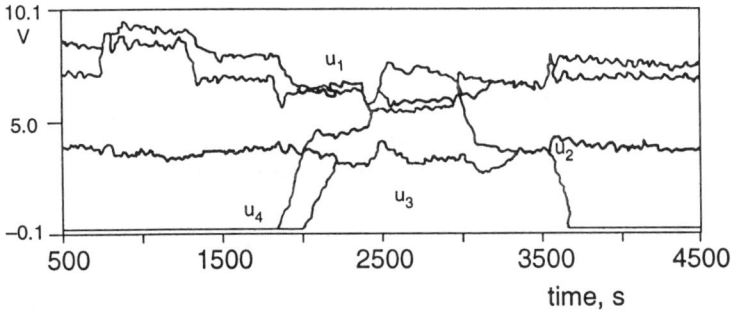

Figure 12.10   Control signals

Some experimental results with optimal boundary control of the heated copper bar, using inversion of the heat equation, are given in Figure 12.11. The aim of the inversion task is to calculate the optimal reference signal $y_r(t)$ for the controller. Following this signal the system reaches desired temperature profiles $s_r(x, t_i)$ in the bar at given time instants t_i. The experiment starts from the known initial temperature distribution in the bar at time instant $t_0 = 0\,$s (see Figure 12.13). The inversion tasks are solved for time instants t_1 and t_2; $t_1 = 840$ s, $t_2 = 1680$ s. Figure 12.12 illustrates the control input signal $u(t)$ to the heater. In Figure 12.11 the real boundary temperature $y(t)$ of the bar and optimal reference signal $y_r(t)$ are depicted. The optimal reference signal is calculated using software for inversion tasks. The aim is to reach, in advance, given temperature profiles $s_r(x, t_i)$ at the time instants t_1 and t_2. The agreement between the desired profiles, $s_r(x, t_1)$, $s_r(x, t_2)$ and real measured profiles can be judged from Figure 12.13. The results obtained show good numerical stability of the proposed inversion algorithm for the cases of inconsistent and consistent tasks.

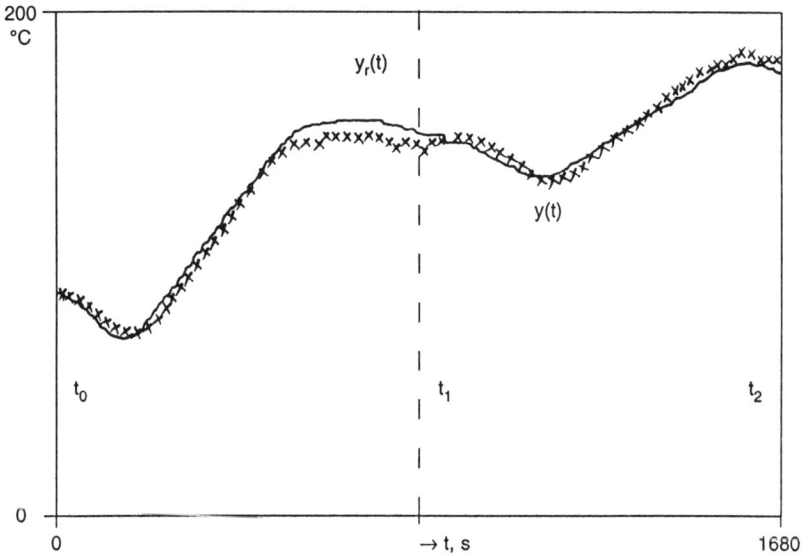

Figure 12.11   Optimal boundary control

        *** $y(t)$ = boundary temperature

        —— $y_r(t)$ = reference value defined by inversion

12.8 Conclusions

Based on a spline approximation for spatially distributed signals and simple system decomposition, a technique for modelling and control of distributed parameter processes has been developed. The technique is applicable for distributed as well as boundary control. A reference signal for boundary control to be tracked by the control system, is generated on the basis of an inversion of the given distributed parameter model. Although the experiments were performed on spatially one-dimensional thermal systems, the ideas can be extended to more complex distributed systems.

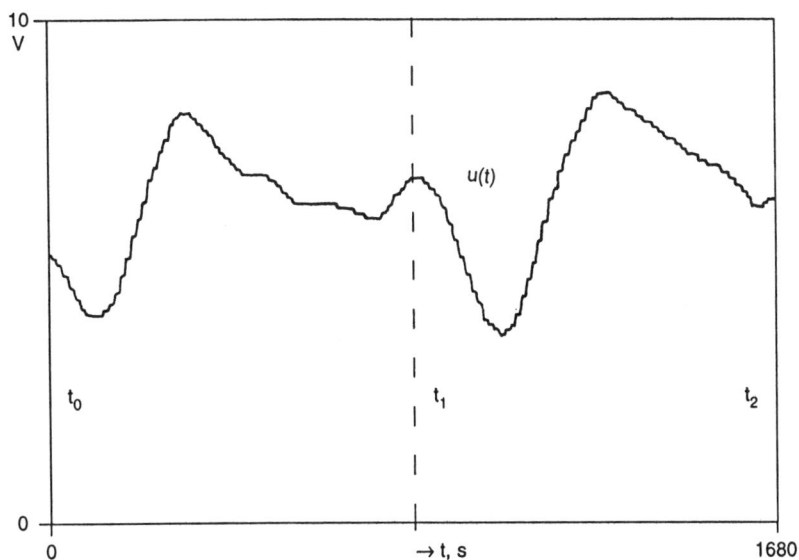

Figure 12.12   Control signal

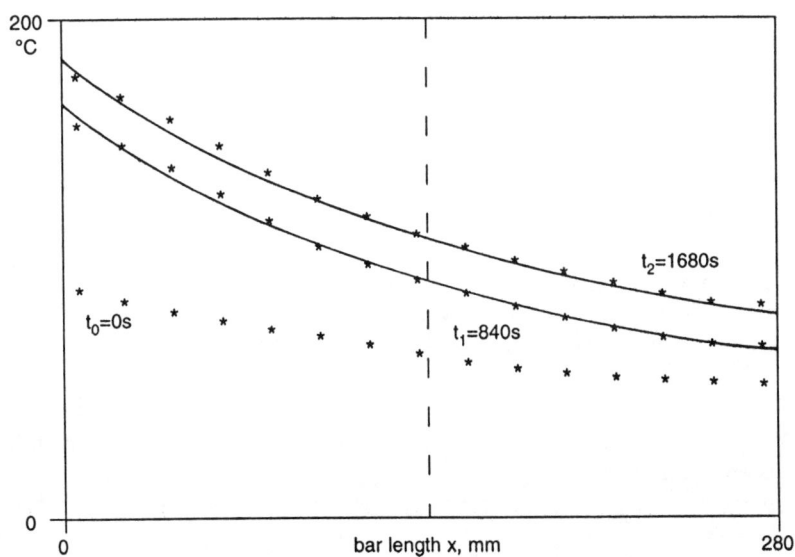

Figure 12.13   The agreement between desired and measured temperature profiles at time instants t_1, t_2

*** $s(x,t_i)$ = measured temperature profile

—— $s_r(x,t_i)$ = desired temperature profile

12.9 References

1   RAY, W. H. and LAINIOTIS, D. G.: 'Distributed parameter systems', (Marcel Dekker, Inc., New York, 1978)

2   TZAFESTAS, S. G.: 'Distributed parameter control systems', (Pergamon Press. Oxford, 1982)

3   SUNAHARA, Y., TZAFESTAS, S. G. and FUTAGAMI, T. (Eds.): 'Modelling and simulation of distributed parameter systems', *Proc. IMACS/IFAC International Symposium*, Hiroshima, Japan, 1987

4   RAUCH, H. E.: (Ed.) 'Control of distributed parameter systems', *Proc. 4th IFAC Symposium I,* Los Angeles, California, USA, 1986

5   BANKS, S. P.: 'State-space and frequency-domain methods in the control of distributed parameter systems', (Peter Peregrinus Ltd., London. UK, 1983)

6   ROHAL-ILKIV, B., ZELINKA, P., RICHTER, R., SROKA, I. and ORSZAGHOVA, Z.: 'Design of multivariable self-tuning controller for a class of distributed parameter systems', in K. WARWICK, M. KARNO, A. HALOUSKOVA (Eds.): *Advanced Methods in Adaptive Control for Industrial Applications*, Springer-Verlag. Berlin - Heidelberg - New York, **158**, (1991) pp.233-250

7   PETERKA, V.: 'Bayesian approach to system identification', 'Trends and progress in system identification', (Pergamon Press, Oxford, 1981)

8   ZELINKA, P.: 'Adaptive control of distributed parameter systems', PhD thesis, Department of automatic control and measurement, Slovak Technical University, Bratislava, Slovakia, (in Slovak), 1989

9   ROHAL-ILKIV, B., ORSZAGHOVA, Z. and HRUZ, T.: 'A stepwise technique for inverse problem in optimal boundary control of thermal systems', in *First International Conference on Inverse Problems in Engineering: Theory and Practice*, Sheraton Palm Coast, FL, USA, 1993

Chapter 13

Fuzzy control: demonstrated with the inverted pendulum

P.M. Frank and N. Kiupel

13.1 Introduction

In this chapter we report on fuzzy logic control and present results of several feasibility studies. In particular we present the inverted pendulum, amongst others, as a practical example. A simple fuzzy controller is designed for this system. Unlike observer-based approaches, no mathematical model is used. The control strategy is generated by formulating the tasks which need to be carried out to keep the pendulum in an upright position. Certain results of the inverted pendulum control system are presented here.

13.2 Control problem formulation

13.2.1 Fuzzy operators

In this section we give a brief description of the fuzzy operators used for this study. For more details see, for example, Reference 7.

As shown in Figure 13.1, the controller consists of the following three parts:

- Fuzzifier

- Inference network

- Defuzzifier

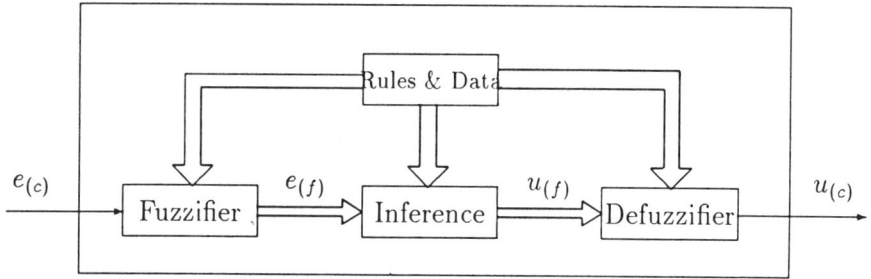

Figure 13.1   Parts of the fuzzy controller

The fuzzifier transforms a crisp value into a fuzzy value. Mathematically this can be denoted as:

$$x \rightarrow a_1 \mu_1(x) + a_2 \mu_2(x) + \ldots + a_n \mu_n(x) \tag{13.1}$$

where n is the number of fuzzy sets influenced by the crisp value x, $x \in [-L, L]$ and $\mu_i(x)$, $i \in [0, 1]$ is the compatibility degree of the ith fuzzy set. The a_i, $a_i \in [0, 1] \forall i$ describe the fuzzification operator which is in this case the product of the compatibility degree and the membership function. The inference network then provides the fuzzy output value $u_{(f)}$. The fuzzy output $u_{(f)}$ is the accumulated output of all the rules in the inference network.

The inference network consists of r rules, where r is:

$$r = r_1 + \ldots + r_k \tag{13.2}$$

r_k is the number of fuzzy sets for each state variable. Each rule in the inference network can be described as follows:

$$(IF \quad u_{(f)i,j} = \mu_{1,1}) \quad AND \quad (IF \quad u_{(f)i,j+1} = \mu_{1,2}) \quad AND \quad (IF\ldots)$$
$$THEN \quad u_{(f)i} = \mu_{oi} \tag{13.3}$$

where μ_{oi} denotes the ith membership function of the output. Then the fuzzy output $u_{(f)}$ is the unification of all $u_{(f)i}$:

$$u_{(f)} = u_{(f)_1} OR \; u_{(f)_2} OR\ldots \tag{13.4}$$

The aggregation operators AND and OR, used in the inference network, can be described in several ways. The minimum operator is used for the AND aggregation operator. Therefore, for the two input values x_1 and x_2, which belong to fuzzy sets A and B, respectively, we obtain the aggregation value:

$$u_{(f)_i} = MIN[\mu_A(x_1), \mu_B(x_2)]$$ (13.5)

The maximum operator is used for the OR aggregation. With the definitions above this results in:

$$u_{(f)} = MAX(u_{(f)_1}, \ldots, u_{(f)_r})$$ (13.6)

The last component is the defuzzifier. The defuzzifier calculates the crisp value $u_{(c)}$ from the fuzzy value $u_{(f)}$. In our case we use the centre of area (COA) method, obtaining the output $u_{(c)}$ from the input $u_{(f)}$ as follows:

$$u_{(c)} = \frac{\int\limits_{-L}^{L} x u_{(f)}(x)dx}{\int\limits_{-L}^{L} u_{(f)}(x)dx}$$ (13.7)

13.3 Technical approaches

Here we wish to present some feasibility studies in order to show that some fuzzy controllers have already been implemented. Therefore three examples have been chosen:

1. A fuzzy controller for a steam turbine;

2. A fuzzy controller for flight control;

3. A fuzzy controller for the inverted pendulum.

The last of these three examples we will explain in more detail, whilst the other two are only briefly summarised.

13.3.1 Fuzzy controller for a steam turbine

In this feasibility study a fuzzy logic speed controller has to be applied to a steam turbine. The efficiency of the fuzzy control was compared with that of conventional PID control. As a basis for this comparison we used a mathematical model of the turbine which has been well validated on a real turbine, in connection with the design of the PID control. A simple fuzzy controller was designed for this model. To assess the resulting fuzzy control we studied by simulation the step response of the speed due to a sudden load reduction of 100 %, as well as the robustness with respect to parameter variations of the turbine.

The control task can be divided into two parts:

1.   Speed control of the turbine;

2.   Pressure control of the turbine.

In the speed control of a steam turbine the following two tasks are of primary interest:

•   Run-up of the turbine to nominal speed;

•   Catch a sudden load reduction at nominal speed.

To investigate the second task, we use both the dynamic model of the steam turbine as well as the PID control concept. It can be seen from Figure 13.2 that the overshoot of the fuzzy-controlled turbine, due to a sudden load reduction of the turbine, is much less than the overshoot of the PID-controlled turbine. The simulation results show that even a relatively simple fuzzy logic control algorithm can produce better results than a PID controller. More details of the control of the turbine can be found in Reference 5.

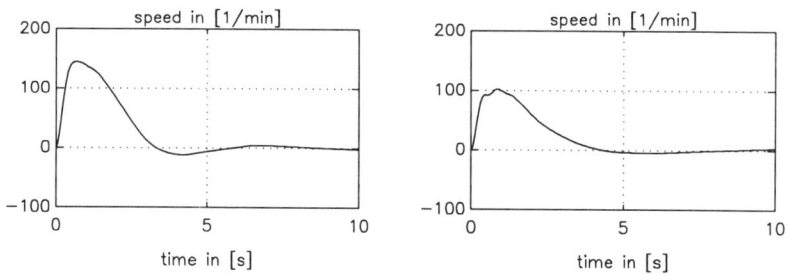

Figure 13.2   *Results of a sudden load reduction of the turbine*
                 (a) PID
                 (b) fuzzy-controlled turbine

13.3.2 Fuzzy controller for flight control

The focus of this feasibility study was the design of a flight control concept using state-space methods. The main concerns for the design were to achieve good stationary accuracy and good robustness. Some of the parameters are highly dependent on the speed of the aircraft, so much so that the aircraft can be unstable under certain parameter changes. As a result the poles of the system are also very strongly dependent on these parameters. Classical controllers solve this problem by changing the parameters according to the speed of the aircraft (gain scheduling).

With the fuzzy controller used the goal can be achieved for the whole speed range without changing any parameters. The only task was to make an appropriate choice of the fuzzy controller parameter. The controller is implemented as a cascade controller, as that reduces the number of combinations of rules significantly.

Even the technical realisation can be done without too much effort in the calculation. In this case a pure software solution was possible. Besides its robustness in the face of parameter variations, the controller is robust against sensor faults too. This indicates that the fuzzy controller is superior to a classical controller. For more details the reader is referred to Reference 6.

13.3.3 Fuzzy controller for the inverted pendulum

The 'inverted pendulum' system has already been described in chapter 7 and, for convenience, is shown again in Figure 13.3

In this case, the following quantities are measured at the pendulum:

1. The position of the cart (x);

2. The angle of the pendulum rod (ϕ).

The control problem can also be solved using classical methods, e.g. state-space feedback, as discussed in chapter 7.

In this section we will solve the control problem using a fuzzy logic controller. Instead of generating a model of the inverted pendulum for control, the so-called 'human-decision-making-procedure' is used. This means that the model is built using a human description of the process rather than by generating differential equations. The basis for the use of such human modelling is fuzzy logic. This means, also, that no mathematical model of the process is needed, unlike in classical control, e.g. an observer-based method, where a mathematical model is required. A block diagram of the controlled inverted pendulum is shown in Figure 13.4.

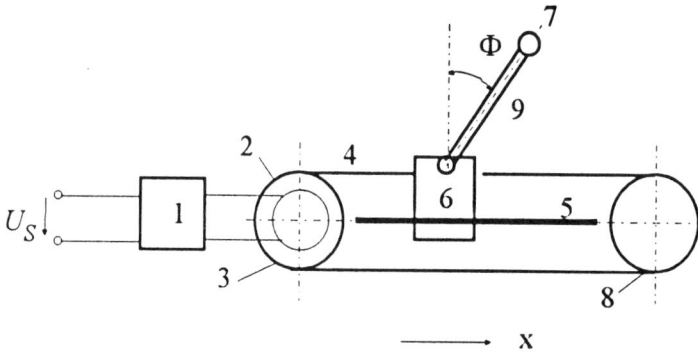

Figure 13.3   *Basic principles of the inverted pendulum*
  1   Servo amplifier
  2   Motor
  3   Drive wheel
  4   Transmission belt
  5   Metal guided bar
  6   Cart

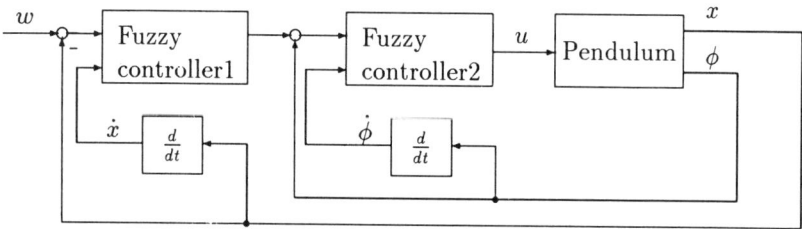

Figure 13.4   *Block diagram of the closed loop with the cascade fuzzy controller*

The two controllers are implemented as cascade controllers. The inner loop controls the angle and the angular velocity of the pendulum while the outer loop controls the position and the velocity of the cart.

Because the only quantities measured are the position and the angle, the derivatives of the position $\dot{x}$ and the angle $\dot{\phi}$, respectively, are generated using the difference between two measurements and normalised with the sampling time. With these derivatives all four state variables are available for control. Now it is necessary to describe the fuzzy controller itself.

The outer loop (position and velocity) is designed as follows:

- 5 fuzzy sets for the position;

- 5 fuzzy sets for the velocity;

- 7 fuzzy sets for the output of this controller.

This results in 5×5=25 rules. Both the fuzzy sets and the fuzzy rules are shown in Figure 13.5.

The inner loop (angle and angular velocity) is then designed as follows:

- 5 fuzzy sets for the angle;

- 5 fuzzy sets for the angular velocity ;

- 7 fuzzy sets for the output of this controller.

This also results in 5×5=25 rules. Both the fuzzy sets and the fuzzy rules are shown in Figure 13.6.

13.4 Discussion

There is substantial freedom in designing the fuzzy controller. Here we are restricted to the controller algorithms described above, but many modifications are possible, such as:

- Increasing the fuzzy sets of the state variables;

- Variation of the shape of the membership functions;

- Introduction of a reliability factor for each rule;

- Using a different method for defuzzification;

- Using different fuzzy operators.

Nevertheless, the structure of the controller is relatively simple and follows directly from the intuition of the design engineer. In our case the design was very straightforward.

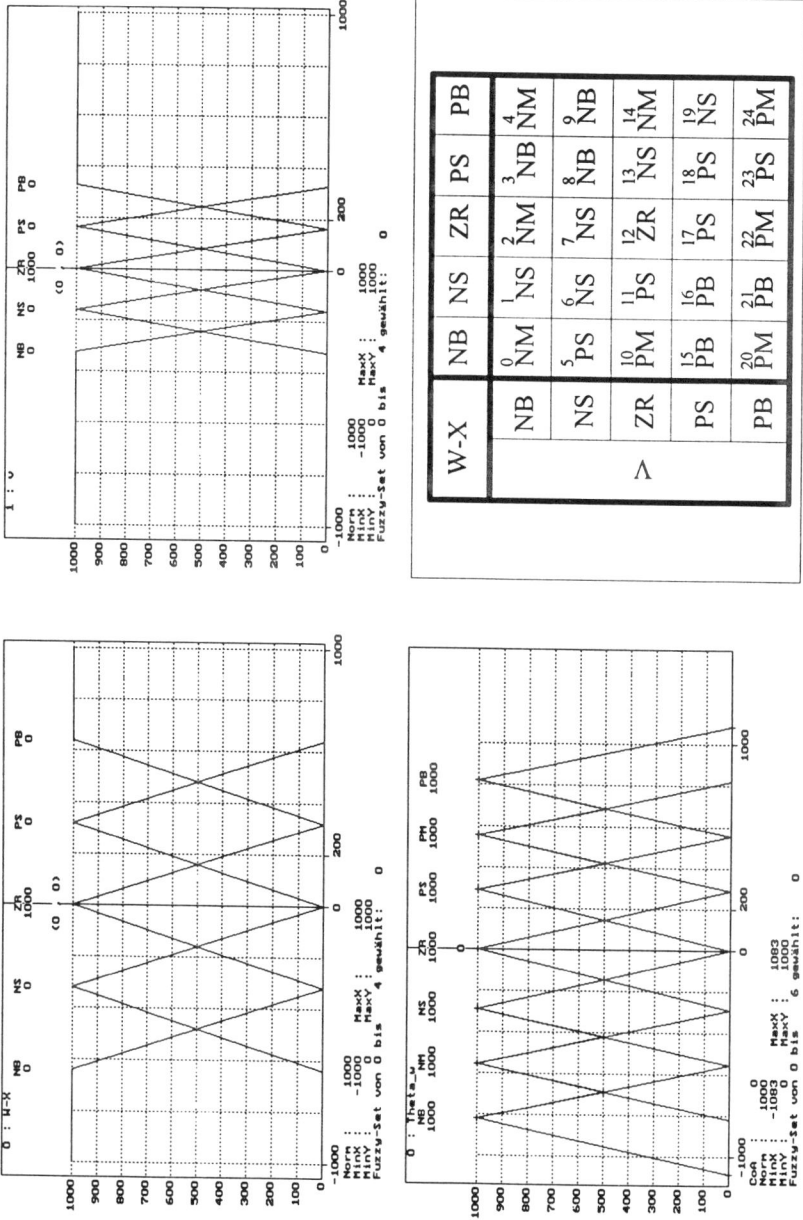

Figure 13.5   Fuzzy sets and rules for the outer cascade loop

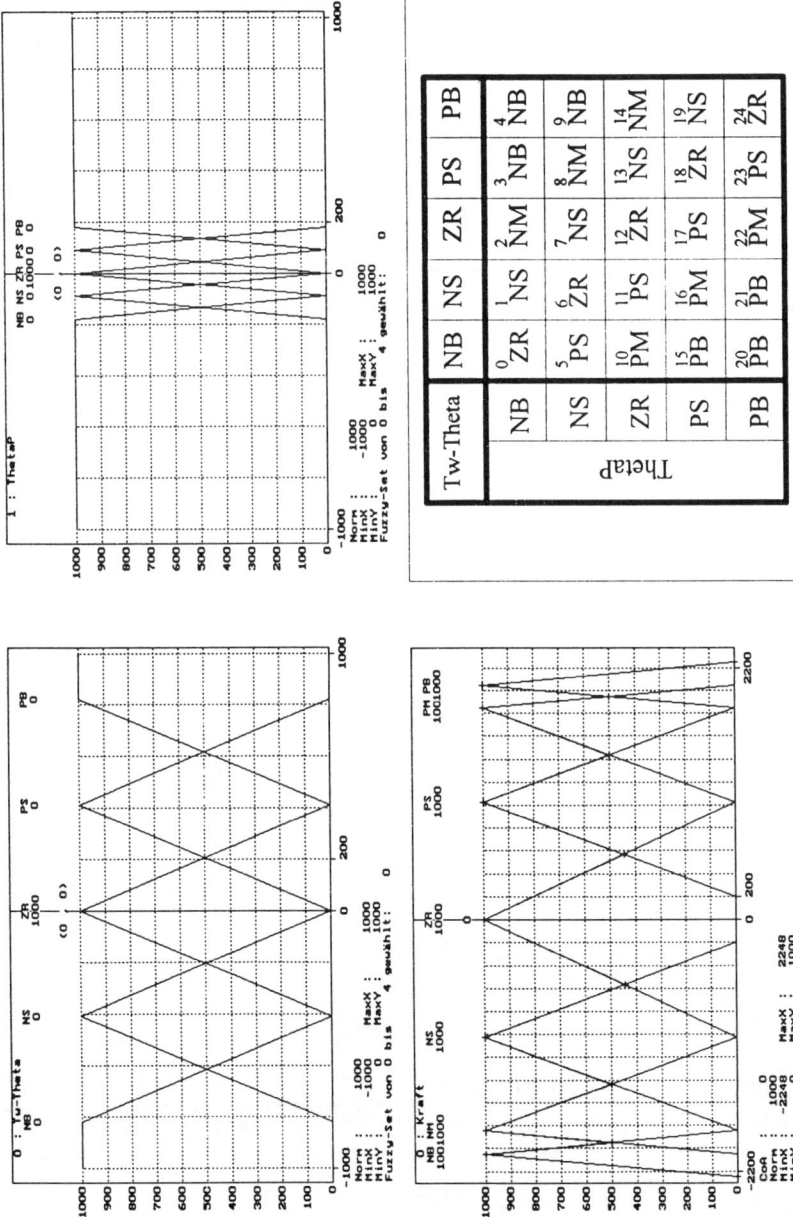

Tw-Theta	NB	NS	ZR	PS	PB
NB	0 ZR	1 NS	2 NM	3 NB	4 NB
NS	5 PS	6 ZR	7 NS	8 NM	9 NB
ZR	10 PM	11 PS	12 ZR	13 NS	14 NM
PS	15 PB	16 PM	17 PS	18 ZR	19 NS
PB	20 PB	21 PB	22 PM	23 PS	24 ZR

Figure 13.6   Fuzzy sets and rules for the inner cascade loop

13.5 Laboratory set-up

The experimental set-up, which is commercially available (see, for instance, Reference 2), consists of the following components:

1. Laboratory model of the inverted pendulum and cart, respectively;

2. Actuator;

3. Computer with terminal;

4. Plotter.

The model of the inverted pendulum can be reduced to the 'cart' system by removing the pendulum rod. This is useful because now the system is simpler with just two state variables $(x, \dot{x})$. For the reduced system, it is even easier to generate a fuzzy controller. It may be helpful to divide the laboratory example into two parts, the cart system for beginners and the inverted pendulum for more advanced students.

13.6 Suggested experiments

At the beginning, one can start with the cart and try to find the fuzzy sets and rules for this plant. There are additional programs which enable one to generate rules and help to implement them for use with the pendulum.

As a next step one can try to add the next cascade to control the upright position of the pendulum rod. This can be done in the same way as the cart control.

To get an idea of the robustness, some variations can be made:

*   Variation of the shape of the membership functions;

*   Variation of the number of membership functions;

*   Variation of the defuzzification method;

*   Variation of the mass of the pendulum rod;

*   Variation of the length of the pendulum rod.

13.7 Illustrative results

Here we want simply to present some results of the fuzzy controller implemented for the inverted pendulum system. Therefore a step response of the position is presented. The results are shown in Figures 13.7 and 13.8, respectively.

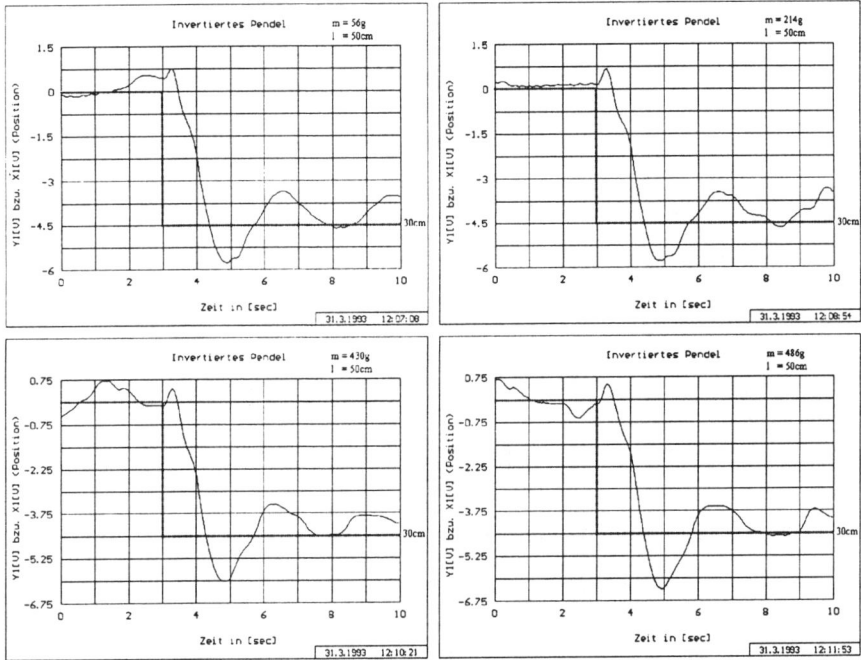

Figure 13.7   Step response with the nominal (long) pendulum rod

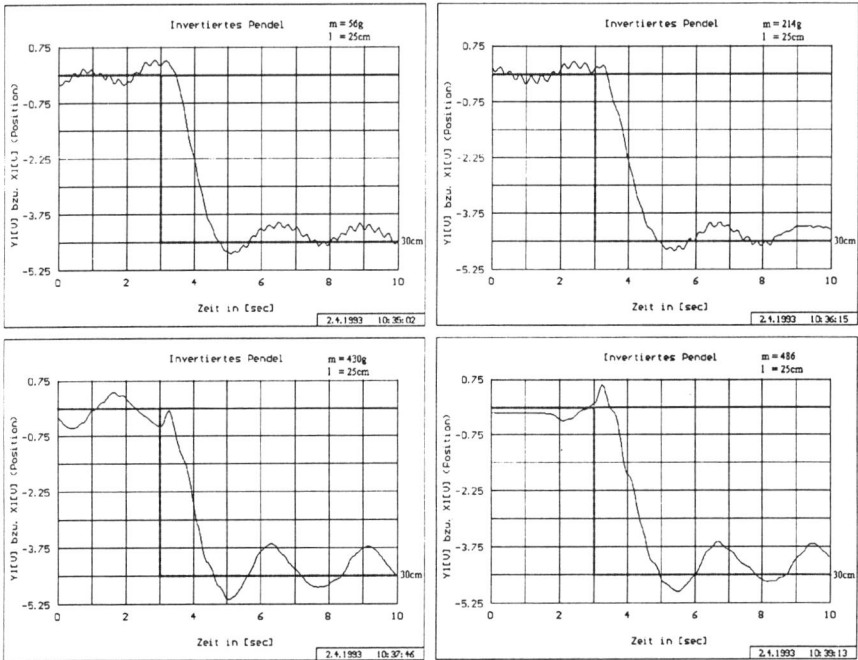

Figure 13.8   Step response with the short pendulum rod

13.8 Conclusion

In this chapter we have looked at fuzzy control of a laboratory model of an inverted pendulum. Both the mathematical models of the inverted pendulum and the fuzzy controller have been described. The task consists of the control of a subsystem of the inverted pendulum, in this case the cart position without the pendulum rod, and of the inverted pendulum itself. The latter is more complicated because it is an open-loop unstable system, while the cart without the pendulum rod is open-loop stable.

This is a good example to teach students how fuzzy logic control can be applied to unstable non-linear systems with non-minimum phase. Control of the inverted pendulum is, for example, similar to altitude control of flight systems and the control of many mechanical systems.

13.9 References

1   BALLAY, R.: 'Real time control of the inverted pendulum using a fuzzy controller, Diplomarbeit, University of Duisberg, Germany, 1993

2   Amira GmbH: Manual for the laboratory set-up of inverted pendulum control, Bismarckstrasse 67, 47048 Duisburg, Germany, 1993

3   OSTERAG, E.: 'Fuzzy Control of an Inverted Pendulum with Fuzzy Compensation of Friction Forces', *Proc. 6th IAR Colloquium 'Fuzzy Signal Processing and Lean Production'*, Duisburg, Germany, 1992

4   BERTRAM, T.: 'Entwurf einer Fuzzy-Regelung am physikalischen Modell eines Drehschwingers', *Proc. 6th IAR Colloquium 'Fuzzy Signal Processing and Lean Production'*, Duisburg, Germany, 1992

5   Entwicklung eines Fuzzy Reglers fur eine Damfturbine, Casestudy, University of Duisburg, May 1993

6   Abschlubericht der Projekstudie 'Fuzzy Logik fuer die Flugkoerperregelung', University of Duisburg, July 1993

7   LEE, C.C.: 'Fuzzy logic in Control Systems: Fuzzy Logic Controller', (1990) IEEE Trans. Syst. Man Cybern., **SMC-20**, No.2, pp. 404-435

Chapter 14

Adaptive control supervision

M. Martínez, P. Albertos, J. Picó and F. Morant

It is known that it is possible to find a suitable combination of control and parameter estimation algorithms in self-tuning regulator (STR) adaptive control, such that the overall system achieves a satisfactory performance, assuming that certain preconditions for stability and convergence of the algorithms are satisfied. This is not always possible, principally in well-tuned closed-loop control, because the control signal is not persistently exciting. Some authors have proposed the development of a higher level of control which has two different goals: (i) to check on-line the aforementioned conditions; and (ii) to take appropriate actions in order to maintain functionality of the overall system when some of these conditions fail. In this work we analyse the information needed at this level in order to take the appropriate action. We apply this methodology to the implementation of a supervisory level for real-time control of coupled-tank equipment.

.

14.1 Control problem

Adaptive control has been studied for some time, and a number of industrial products can be found in the market (e.g. Novatune, Elextromax V, Exact, etc.) which are used in many industrial adaptive control loops. These involve a wide range of applications in aerospace, process control, ship steering, robotics and other industrial control systems. Chapter 12 of [1] presents an extensive number of applications of adaptive control.

These applications have shown that there are many cases in which adaptive control is very useful, others in which the benefits are marginal and others in which control is impossible. In this last case, the reason is often the presence of problems concerning critical processes [1,2].

Pure adaptive control systems may be insufficient when we are controlling processes, especially in real plant (industrial plant or laboratory equipment). The fundamental reason for this insufficiency is violation of certain conditions for correct function of the overall system, either from the start-up or during normal operation [3–5].

Possible violations of these pre-conditions have been classified [5–7] including:

- incorrect choice of process structure (order and/or delay);
- non-linearities;
- deficient signal excitation;
- presence of unmodelled dynamics;
- process parameter variations;
- inadequate sampling time ;
- unstable poles and zeros;
- imperfect cancellation (critical for some designs); and
- loop instability.

These problems can be overcome through appropriate selection of algorithms and tuning parameters, selections based mostly on previous knowledge and experience.

In an adaptive scheme the following basic blocks may be considered:

- **Parameter estimation module:** in this module we seek the most appropriate parameters, within a family of models, to fit the process' dynamic behaviour;

- **Controller design module:** according to the stated objectives and the process model previously obtained, at least one control law is proposed. Usually, the control solution is not unique and additional constraints, goals or requirements will determine which one is the most suitable;

- **Closed loop evaluation module:** this module verifies the fulfilment of control specifications, looking for abnormal situations, such as bumping, hidden oscillations, long term drift etc.

At the lower levels in the hierarchy the problems can be handled algorithmically. A number of indicators, such as prediction error, trace of estimated covariance matrix, pole location and cost indices, amongst many others, may be evaluated. These indicators, together with some experimental knowledge, should be used at the top level, the supervisor level. In the following we are going to develop a supervisor level which can be designed using AI techniques. The basic scheme of such a supervisory system is depicted in Figure 14.1.

The concept of an index or indicator, as the element capable of detecting the aforementioned violations, has then been introduced [5] and justified because of the impossibility of getting general solutions to the problems of stability and robustness of adaptive algorithms.

Furthermore, the special characteristics of real processes, with the presence of uncertainty and imprecise knowledge, make so-called heuristic indicators [5] especially useful, due to the ease of interpreting high-level qualitative aspects within the supervision methodology.

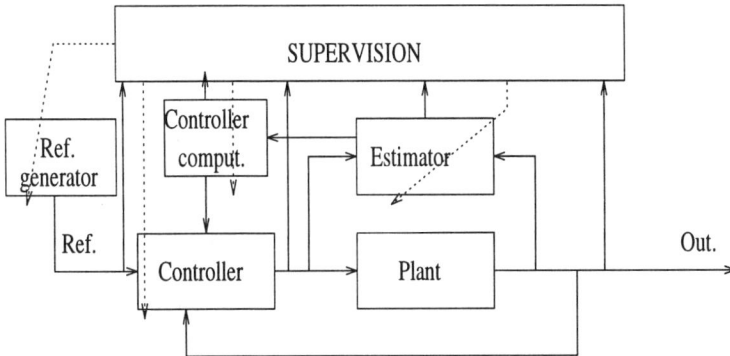

Figure 14.1 Supervision system

14.2 Technical background

14.2.1 Supervision tasks

In this section we will introduce the main problems found in an adaptive system and the tasks to be undertaken at the supervisory level. We will focus our attention on explicit self-tuning regulators, since they are widely used, and in them we can see most of the problems which arise in any adaptive scheme. The basic tasks performed by the supervisor will be shown and the concept of supervisor indicators introduced.

Supervisor tasks can be grouped into four main sets [5] relating to the part of the system they effect:

(1)   Start-up or pre-identification;
(2)   Parameter estimation;
(3)   Controller synthesis;
(4)   Loop supervision.

14.2.1.1   Start-up or pre-identification

Strictly speaking, this is not a supervisory task, but it will produce some design factors which will be supervised on-line (for instance, the model order, delay estimates and disturbance model structure), or will be used by the supervisor. If the adaptive controller was applied from the very beginning, we would face the problem of a lack of estimates of the process parameters, which in turn would produce undetermined and possibly unacceptable control actions. Therefore, a prior estimation of the process model order, delay and parameters is made by open-loop testing.

Model order estimation may be achieved by several methods [5]; for example, the cost function test. Determination of the model delay is usually carried out either by the cost function test or by using a model with an expanded numerator, the delay being the number of terms in the numerator which can be neglected [5].

Once the process model structure has been determined, the disturbance model structure must be tested. The noise auto-correlation vector is calculated, the noise being estimated as the prediction error in the identification procedure. If the disturbance present in the process can be modelled as white noise, then:

$$\begin{aligned} corr\big[e(k),e(k)\big] &= \sigma_e^2 \neq 0 \\ corr\big[e(k),e(k+\tau)\big] &\approx 0 \quad ; \; \tau > 0 \end{aligned} \qquad (14.1)$$

If the noise is coloured, there will be ν terms in the second expression which are non-zero, ν being the order of the coloured noise which models the process disturbance [8].

14.2.1.2   Parameter estimation

Estimation methods, as those described in chapter 1, constitute the heart of the self-tuning regulator. The reliability of the controller relies to a great extent on correct estimation of parameters (equations 1.1-1.3) [7]. Aspects to be considered include the following.

Signal and estimate filtering.   Input and output signals must be filtered so that the estimation does not become corrupted with noise.

For instance, to avoid corruption of the estimation by abrupt disturbances, the expected process output may be used. The process output and the (current) model output for the (current) input are compared. If the difference appears to be unacceptable, the expected output can be used instead of the actual output. Obviously, a sudden change at the process output may have been originated from a real sudden change in the process parameters, with the process parameter estimates not yet having converged to the correct values. The source of the difference should be detected or, as is typical with most supervision tasks, a trade-off must be established between the rejection of sudden abrupt output signal changes and the estimation reaction speed following sudden changes in the process parameters.

Estimator suitability to signal excitability.   Persistent loop excitability is needed for correct estimation. This persistence is guaranteed by a control signal which sufficiently excites all the modes of the dynamic process. If this condition is not met, for instance, when the controller is well tuned, linearly dependent rows appear in the estimator variance-covariance matrix. Thus, this matrix will become singular or nearly singular in practice, and the estimator equations will become ill-conditioned.

When forgetting factors are used, this problem can be seen from another point of view. From the general RLS estimation equations (equations 1.7-1.9 repeated here for convenience):

$$\theta(k+1) = \theta(k) + K(k+1)\left[y(k+1) - \alpha(k+1)^T\theta(k)\right] \tag{14.2}$$

$$K(k+1) = \frac{P(k+1)\alpha(k+1)}{\lambda + \alpha(k+1)^T P(k)\alpha(k+1)} \tag{14.3}$$

$$P(k+1) = \frac{1}{\lambda}\left[P(k) - \frac{P(k)\alpha(k+1)\alpha(k+1)^T P(k)}{\lambda + \alpha(k+1)^T P(k)\alpha(k+1)}\right] \tag{14.4}$$

Consider the inverse variance-covariance matrix *P(k)*. If no new significant information enters the system, the regression vector will contain negligible terms, and:

$$P(k+1) \approx \frac{P(k)}{\lambda} \tag{14.5}$$

Now P will grow exponentially if $\lambda < 1$, and the parameter estimates $\theta(k)$ will have large excursions – a phenomenon known as *bursting*. So, if no new significant information is present in the data we should stop the estimation.

Algorithm exhaustion.   If $\lambda \approx 1$, since the second term of the right hand side of Equation 14.4 is quadratic, then when the system is properly excited, the values of the P elements will decrease. This means that the estimation algorithm will collapse and so further time-varying parameter estimation becomes impossible. For the correct tracking of variable process parameters, a minimum estimation energy must be kept in the algorithm.

This effect may be monitored through the P matrix trace. When its value falls below a certain threshold, two basic actions can be taken:

(1) Matrix P reinitialisation.   This solution has the disadvantage of modifying the algorithm search direction with corresponding transient behaviour.

(2) Forgetting factor modification.   This modification can be progressive and so does not produce rough transients. If the trace of P is too high with a corresponding change of parameters bursting, the forgetting factor λ should be increased. Conversely, decreasing λ increases the sensitivity of the algorithm to new data. One related possibility is the use of constant trace estimation algorithms [3].

14.2.1.3   Controller design
The goal here is to avoid the design of a new controller (either by changing its parameter or its structure, including change of controller type), as this may degrade the effectiveness of the control.
    Among the functions associated with this supervision task we can find:

Poles and zeroes test.   Both process and disturbance model poles and zeroes are calculated. Depending on these values, different control strategies will be applied. For instance, minimum variance controllers with control action weighting must be used if the process is minimum phase, etc. [9].

Stability test.   Before updating the controller parameters, the closed-loop characteristic equation should be solved in order to determine the stability of the system. Parameters such as the control action weighting in minimum variance controllers can be modified accordingly by the supervisor [5].

14.2.1.4   Loop supervision
The aim of this task is the detection of instabilities in the process input and output signals. If required, loop control is transferred to the back-up controller.

14.2.2 Selection of indicators for supervision

The supervisor must take the signals coming from a control loop (reference, output, control actions) and calculate the indicators associated for adaptive control. These indices or indicators may be associated in groups related to:

(1)   Start-up procedure: pre-identification and model verification
  • Sampling time test
  • On-line estimation of order
  • On-line estimation of delay
  • Unmodelled dynamics test

(2)   Parameter estimation
  - Prediction error indicators
  - Variance-covariance matrix indicators
  - Estimated parameters indicators

(3)   Controller design
  - Indicators for the poles and zeroes of the model
  - Loop error indicators

(4)   Closed-loop
  - Output signal indicators
  - Characteristic equation indicators

Each one of these groups incorporates statistical indicators (means, variances, covariances, norms, trends, accelerations, etc.), and it is possible to establish which of these gives more information about the problems arising during control. A description of numeric indicators is presented in the Appendix to this chapter.

14.2.2.1   Evaluation of indicators

Validation of the indicators can be studied through a simulation package, thus allowing for different estimation and control combinations in open- and closed-loop situations.

Let us review some results that allow us to connect the numeric information with heuristic information. A complete study of these relationships is given in References 8 and 10.

Unmodelled dynamics.   Prediction error autocorrelation is used as an indicator for detecting the presence of unmodelled dynamics.

As is well-known, if the prediction error can be modelled as a white noise process, the autocorrelation function will be zero for those signals shifted by one or more sampling periods. If it can be modelled as a coloured noise process, as many factors as the order of the corresponding filter will be different from zero (Figures 14.2 and 14.3).

Process parameter change.   In order to detect changes in the process parameters, the cross-correlated factors between the prediction error and the control action and process output signals, respectively, can be used.

These indices inform us quickly, not only about generic parameter changes, but also (with some assumptions) about which specific parameter has changed. This is very important, since it enables specific actions to be taken for the parameter or group of parameters under change, leaving the others untouched.

When the estimation is undertaken in the closed loop, the information given by these indicators is not as clear as in the open loop. This is especially true for the process denominator coefficients, due to the presence of feedback (the output is an implicit and explicit function of the parameters). This makes it difficult to specify which parameter is changing. Yet, from a qualitative point of view, the same imprecision may be accepted and it is possible to handle the information given by these indicators (Figure 14.4).

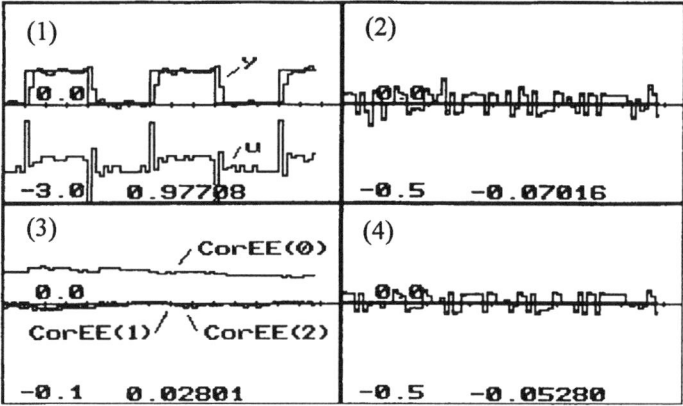

Figure 14.2   Unmodelled dynamics structure. Second-order overdamped process. RLS. (1) Ref.,input,output. (2) A priori prediction error. (3) Autocorrelation $\Phi_{ee}(\tau)$ coefficients. (4) A posteriori prediction error. The only autocorrelation coefficient not zero is $\Phi_{ee}(0)$.

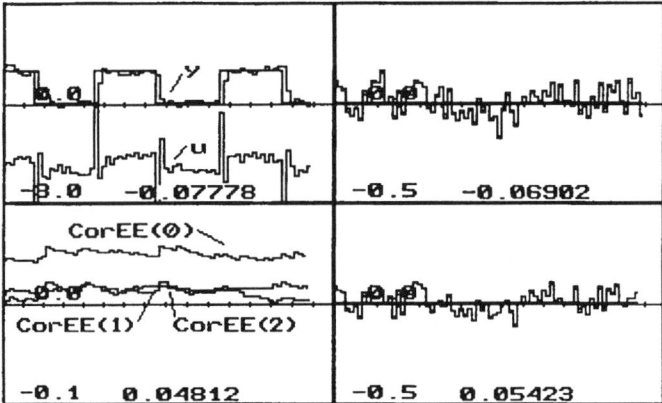

Figure 14.3   A similar process with a coloured disturbance

14.2.2.2 Minimum set
The selection of a minimum set of indicators that represent the process state depends on the individual viewpoint and on the process to be controlled.

Numeric and heuristic indicators must be studied in the evaluation step and a relationship between both kinds of indicators has to be established [8,10]. In Table 14.1, we can see that certain numeric indicators supply similar information about the same heuristic aspects whereas others supply complementary information. We propose a minimum set of indicators which is complementary to the set proposed by Reference 5.

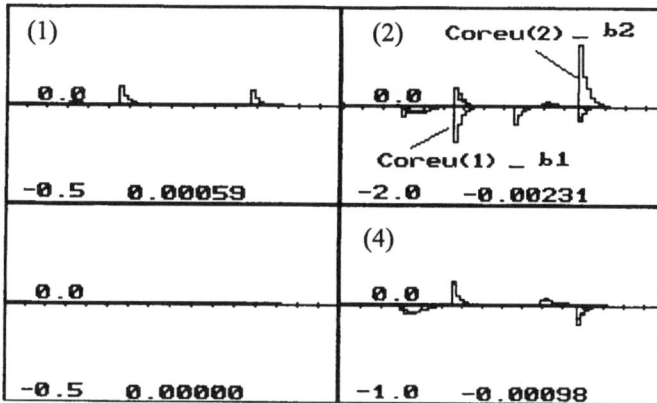

Figure 14.4   Cross-correlation factor evolution when b_1 and b_2 change.
(1) $\Phi_{ee}(\tau)$ (2) $\Phi_{eu}(\tau)$ (4) $\Phi_{ey}(\tau)$.

Heuristic indicator	Numerical indicator
Validity of the process order	Prediction error index
Validity of the process delay	Delay test
Parameters process change	Norm and trend of the parameters
Zeroes process change	Cross-corr. Func. pred-error/input
Poles process change	Cross-corr. Func. pred-error/output
Convergence of the parameters	Variance of the prediction error
Estimation bias	Correlation funct. of the pred. error
Bursting of parameters	Matrix P trace
Energy of the algorithm	Trace of the matrix P
Signal excitability	Trend of the matrix P
Unmodelled dynamics	Variance of the parameters
Admissible set of regulators	Characteristic equation roots
Closed loop stability	Variance of the output

Table 14.1   Minimum set of indicators

14.2.3 Functions and tasks of the supervisory level

Once the malfunction has been detected by the indicators (either individually or together), we need to correlate their responses with some function which takes action in order to correct the problem.

In practice these supervision functions can be classified in differentiated groups according to their actions [8]:

(1)   Updating of filtering factors
  - of signals (inputs and outputs)
  - of parameters

(2)   Fitting process structure changes
  - increment/decrement of model delay
  - increment/decrement of model order
  - change of estimation algorithm

(3)   Handling of process parameter changes
  - variable forgetting factor
  - estimation algorithm re-initialisation

(4)   Ensuring identifiability conditions
  - fixed forgetting factor
  - estimator switch on/off
  - extra signal addition

(5)   Controller choice/tuning
  - choice of compatible controllers
  - scheduling of the design factors
  - change to back-up controller

(6)   Ensuring loop stability
  - robust design
  - back-up controller change

Each supervisory function is the response of the supervisory system to an anomaly which may appear during normal operating conditions and is detected by the indices. One single index, or some combination of several, can detect the anomaly. The information in the set of indices determines the state of the overall system and the most appropriate functions are activated accordingly.

Supervisory functions have been studied, together with suitable indicators, each supervisory function being associated with a state. For instance:

(1)   **State:** process structure change

- **Indices**
  - cancellation of poles and zeroes
  - prediction errors
  - test for time delay detection

- **Function**
  - model structure change
  - controller structure change

(2)   **State**: parameter process change

- **Indices**
  - prediction error correlation
  - norm of estimated parameters
  - variance of estimated parameters

- **Function**
  - forgetting factor modification
  - estimation algorithm start
  - filtering of parameters

(3)   **State**: signal excitation

- **Indices**
  - variance-covariance matrix trace
  - variance-covariance matrix trace trend and acceleration

- **Function**
  - forgetting factor modification
  - extra control signal

The indicator which is associated most suitably with a supervisory function depends on the process to be controlled as well as the environment in which the overall system must operate.

14.2.4 Implementation of supervision functions

14.2.4.1 A classical approach

It is possible to implement a supervisor using modern flexible languages, which feature a set of additional capabilities (real-time programming, easy hardware-level programming, graphic capabilities, etc.), making experimentation possible with the developed algorithms in a real-time environment and on real processes. Furthermore, it is possible to compare the efficiency of these algorithms when distributed on different computers linked by suitable communications.

A tendency exists to implement supervisors in the form of a unique procedure, in which all supervision functions are grouped together. This usually allows for easier programming, and yet experience demonstrates that some supervisory functions should be run at specific points within the control loop with immediate priority. The scheme described above hardly conforms to this kind of strategy. Hence, it seems more logical to design the algorithm that embodies the supervision functions in such a way that it can stand alone. Grouping it (generally in the form of concurrent or parallel processes) could then satisfy the temporal limit within which each function must respond.

14.2.4.2 An artificial intelligence approach

Numeric information can be converted into heuristics using techniques such as fuzzy theory. If we now consider the natural presence of heuristics in control, especially in industrial environments, it is reasonable to conclude that artificial intelligence tools, and particularly expert systems (ES), will be suitable to handle this information and to solve problems of supervision. However, many problems remain to be solved even after taking this approach. Expert systems have mainly been developed for off-line applications, where time is not a critical factor, but now unlikely the situation is that of real-time systems. An additional problem emerges with data consistency when examined by an ES [4].

We find that programming languages dealing with ES are oriented towards conventional problems, tend not to be very versatile and possess very limited mathematical capability. For this reason additional programming effort is necessary to obtain ES environments using general purpose languages. In spite of these difficulties, we think that this is a sound line of research because an ES (with its constituent elements of knowledge base, inference engine and database) appears to be the best way to model the large quantity of imprecise knowledge present in the industrial processes to be controlled [11].

There exists a problem of numeric to heuristic conversion. ESs deal with qualitative heuristic information and therefore we must regard the different supervision indicators in an heuristic fashion. However, data generated in a control loop are basically of numeric nature; we look to the techniques of fuzzy logic to make the translations between the two contexts.

In this sense, we can define a fuzzy controller as a real-time ES architecture that uses fuzzy logic techniques of knowledge representation to define generalised control strategies [8].

14.3 Laboratory set-up

The laboratory set-up is demonstrated using a coupled-tank equipment. This is an example of a MIMO and non-linear process. The equipment has two coupled tanks connected by a variable orifice. The variables to be controlled are the level of both tanks (H_1, H_2) whilst the manipulated variables are the water flows that feed each tank (Q_i, Q_j). Figure 14.5 shows a schematic diagram of this process.

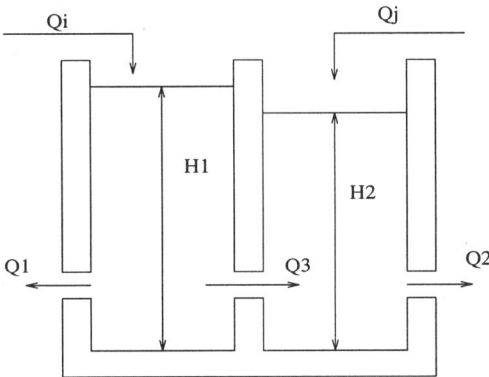

Figure 14.5   Scheme of the coupled-tank equipment

Both tanks are fed, via a safety valve, from a small reservoir with two suction pumps. The flow into each tank can be controlled by modifying the voltage of these pumps. Each tank has a level sensor at the top, to measure the liquid level. The system allows several different configurations depending on the opening of the valves as well as the diameter of the variable orifice between the tanks.

It is possible to determine a mathematical model for this process but, due to physical parameter variations (valves opening, orifice between tanks) and level set-point variations that exist in actual operation, it is preferable to find an experimental model by discrete parameter estimation and to use closed-loop control by STR (self-tuning regulator) adaptive control supervision.

A set of instrumentation is required for digital control of the whole system; some of this has been developed in our laboratory and some purchased. This set of equipment comprises the following:

(1)   **Level transducer**. This transducer has been developed using an ultrasonic transmitter and a receiver.

(2)   The **actuators** are two pumps. In order to control the flow through the pumps it is necessary to modify the input voltage for each pump. This is achieved using a special circuit with an output range between 0 to 10 V.

(3)   **Level interface.** The information from the level sensor is translated into digital information using an interface between the sensor and the computer. The digital signal is obtained by measuring the interval of time between the emission and reception of one ultrasonic wave train.

(4)   A **pump driver.** In order to control the flow through the pump it is necessary to modify the input voltage for each pump. This task is achieved using a special drive with an output range between 0 to 10 V.

(5)   A DT2808 or PCLAB converter with A/D, D/A converter and digital ports for I/O.

(6)   A 486 PC-AT computer or above with mathematical co-processor.

(7)   A package of software including MS-DOS and the Modula-2 compiler.

A general diagram of the control equipment is shown in Figure 14.6.

The signals with the information concerning the levels are translated using a special interface into a digital word which is read into a digital port installed in the computer. The computer works with this information and, depending on the control algorithm, provides control via an analogue signal to vary the flow through the pumps.

The coupled-tank process has the following configuration:

- valve 1 closed / valve 2 open
- pump 1 open / pump 2 open
- controlled variable H_1
- control variable Q_i
- variable orifice 1 or 1.3 cm
- sample period 0.3 s

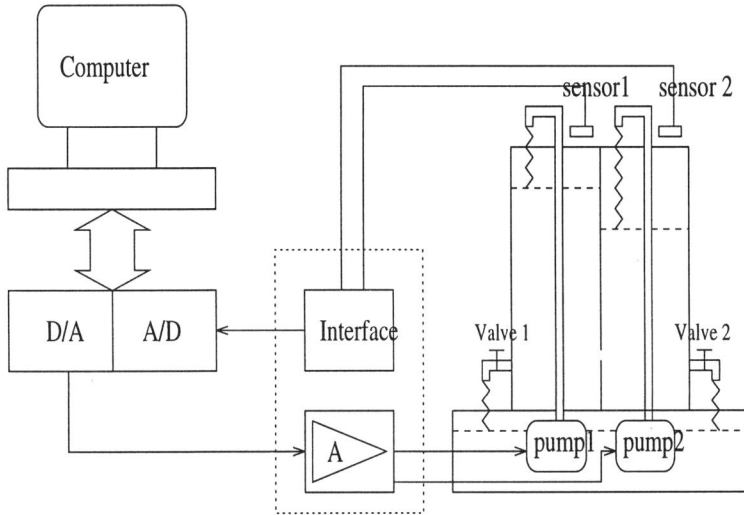

Figure 14.6   *General diagram of the control equipment*

With this configuration we pre-estimate a second-order, linear discrete SISO model using a RLS estimator with a white noise input, defined by the difference equation:

$$A(z^{-1})h_1(k) = B(z^{-1})q_i(k) + v(k) \qquad (14.6)$$

where $h_1(k)$, $q_i(k)$ and $v(k)$ are the input, output and disturbances at instant k and A and B are polynomials defined by:

$$A(z^{-1}) = 1 - 1.036z^{-1} + 0.263z^{-2} \qquad (14.7)$$
$$B(z^{-1}) = 0.1387z^{-1} + 0.0889z^{-2} \qquad (14.8)$$

The disturbances are described by the stochastic process:

$$v(k) = H(z^{-1})e(k) \qquad (14.9)$$

where $H(z^{-1})$ is a rational function, and $\{e(k)\}$ is a white noise process. In this practical case it is possible to assume $H(z^{-1}) = 1$, then the RLS algorithm gives unbiased estimation and $\{e(k)\}$ represents the prediction error.

A simulation package allows more exhaustive experimentation with more complicated plant and stochastic process models $H(z)$.

Many methods can be used for the recursive estimation of the model parameters – RLS, RELS, RIV, RML, STA etc. [6] – the choice of which depends on the nature of $H(z)$.

14.4 Suggested experiments

Experience obtained working with adaptive control in industrial environments highlights certain practical aspects concerning the implementation of this kind of control. We have analysed the most appropriate index set to carry out the supervisory functions.

Based on this idea, an expert system prototype for real-time control, which supervises the correct behaviour of the adaptive loop, can be developed. The use of the aforementioned process simulator allows for validation and adjustment of the expert system. Some of the most useful results that are obtained follow here.

14.4.1 Forgetting factor scheduling

The fuzzy controller developed changes the value of the estimation forgetting factor to cope with different specified events. These events are detected by the heuristic indicators (parameter blow-up, algorithm extinction, lack of excitation, parameter change, or parameter convergence) which are then used by a set of rules in the fuzzy controller. The state of the heuristic indicators is determined by the corresponding numeric indicators in a monitoring table, translated through an event table using fuzzy logic.

An example of closed-loop control with the coupled tanks is presented in Figure 14.7. In the upper left part can be seen the controlled variable $(y=h_1)$ and its set-point. The fuzzy controller detects two changes in the excitability conditions at the beginning of the experiment, and it decreases the forgetting factor (f.f. in the upper right part) with the intention to speed up forgetting of the old parameters. When the excitation falls, the fuzzy controller increases the f.f. with the intention of preventing the parameters from blowing up. In the final situation the matrix P trace (trP in the lower left part) is an indicator that illustrates poor excitation in the estimation algorithms.

14.4.2 Estimator scheduling

A second closed-loop experiment using the coupled tanks is presented in Figure 14.8. The fuzzy controller tries to detect an unsuitable choice of estimator type. Two estimators with wide applicability – a RLS and a RELS – run in parallel, and rules based on the corresponding autocorrelation of the prediction errors are used to change the estimator.

At the beginning, the RLS algorithm is active whilst the RELS is a back-up algorithm. In the figure the lower diagrams present the first two elements of the prediction error autocorrelation vector ($\Phi_{ee}(0)$ up and $\Phi_{ee}(1)$ down), RLS on the right, and RELS on the left.

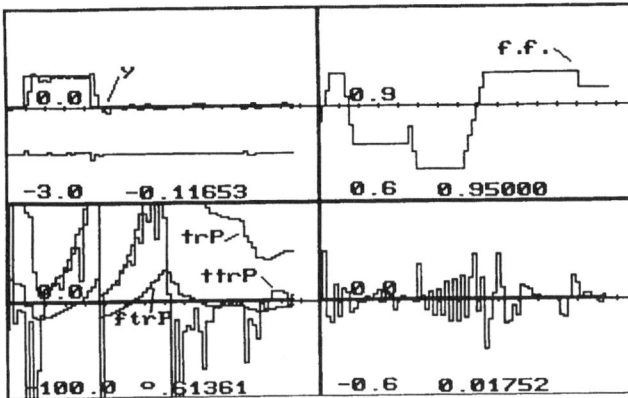

Figure 14.7   Forgetting factor scheduling

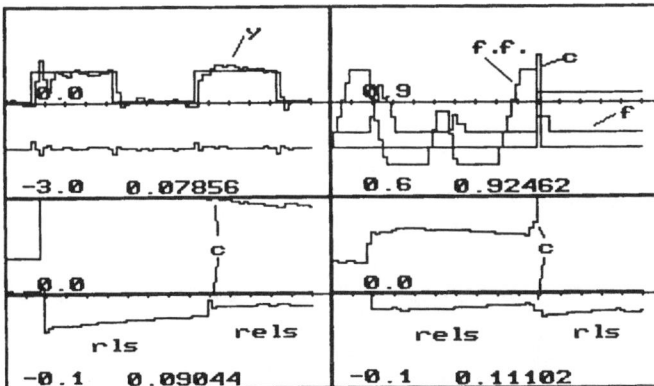

Figure 14.8   Estimator scheduling

The controlled variable y incorporates a coloured noise term which is not detected by the fuzzy controller in the first step of the experiment. Thus, in this step the controlled variable follows the set-point poorly (first step in upper left diagram). When the excitation in the estimation algorithms (RLS and RELS) modifies both autocorrelation vectors Φ_{eeRLS} and Φ_{eeRELS}, then $\Phi_{eeRELS} < \Phi_{eeRLS}$

and, at this moment, the fuzzy controller decides to change the estimator type, from RLS to RELS (warning with the letter c in the upper right diagram). After this, the controlled variable behaviour is improved.

14.5 Conclusions

The experience obtained working with adaptive control in real processes has some practical benefits in dealing with the implementation of this kind of control. In this chapter, relevant issues in adaptive control supervision have been highlighted to draw attention to the difficulties of adaptive control implementation in industrial environments. Based on this idea, a fuzzy controller supervising the correct behaviour of the adaptive loop has also been included.

A laboratory set-up based on a configurable coupled-tank process has been used. In one of these configurations, we have chosen the more appropriate indicator set to carry out the supervisory functions, and two of the more relevant results, forgetting factor scheduling and estimator scheduling, have been presented.

The tool we have developed is simple to handle and very suitable for training students, allowing for a number of process configurations with different control problems. Further, by means of a simulation package for discrete models, more exhaustive experimentation with other models – which are more complicated in dynamic terms than the physical coupled-tank plant – is available.

14.6 References

1   ASTRÖM, K. J. and WITTENMARK, B. 'Adaptive Control'. Addison-Wesley publishing Company, 1989

2   ALBERTOS, P., NAVARRO, J.L., MARTINEZ, M., and MORANT, F. 'Intelligent Industrial Control'. 11th World Congress IFAC. Tallin, URSS, 1990

3   FORTESCUE, T. L., KERSHENBAUM, L.S., and YDSITIE, B.E. 'Implementation of Self Tuning Regulator with Variable Forgetting Factors'. *Automatica*, 1981, **17,** No. 6, pp.831-885

4   ISERMANN R. 'Parameter Adaptive Control Algorithms-A tutorial'. *Automatica*, 1982, **18**, pp.513-528

5   ISERMANN R., LACHMANN, R. 'Parameter-Adaptive Control with Configuration Aids and Supervision Function'. *Automatica*, 1985, **21**, No. 6, pp.625-638

6   LJUNG L., SODESTRÖM T. 'Theory and Practice of Recursive Identification'. MIT Press, 1983

7   SCHUMANN, R. 'Towards Applicability of Parameter Adaptive Control Algorithms '. Proc. IFAC Congress Kyoto, Pergamon Press, 1981

8   MARTINEZ, M. 'Indicators for Adaptive Control Supervision. Implementation by means of Expert System Methodology'. PhD Thesis Universidad Politécnica de Valencia (Spain), 1991

9   ISERMANN R. 'Digital Control Systems'. Springer-Verlag NY, 1981.

10   MARTINEZ, M., MORANT, F., and PICO, J. 'Supervised Adaptive Control'. *Applications of Artificial Intelligence in Process Control*. Ed. Pergamon Press. ISBN 0-08-042016-8. pp.427-455, 1992

11   MORANT, F., MARTINEZ, M. 'Hierarchical Expert System as Supervisory Level in an Adaptive Control'. 4th IEEE I. Symp. on Intelligent Control. Albany, September 1989

14.7   Appendix: numeric indicators

14.7.1   Associated with the pre-identification process

The pre-identification process is responsible for producing an adequate start-up in the adaptive control system. Therefore, it must provide enough information to avoid violating the system pre-conditions. During pre-identification, which is carried out in open-loop, the initial design factors for the on-line control start-up will be proposed from the associated numeric indicators.

Indicators

(1)   **The sampling rate test** allows for the most appropriate sampling rate to be chosen from the determination of the 95% settling time [5].

$$\frac{1}{10}T_{95} \leq T_0 \leq \frac{1}{4}T_{95} \qquad (14.10)$$

(2)   **The process order test.** The prediction error quadratic function may be used to suggest the process model of order m:

$$V(m) = e(m)^T e(m) \qquad (14.11)$$

Alternative tests, such as the poles-zeroes test or the residues test exist, but they are more time- and memory-consuming.

(3)   **The process delay test** is based on the B polynomial coefficients which can be neglected:

$$\left|\hat{b}_1\right|, \left|\hat{b}_2\right|, \ldots, \left|\hat{b}_\beta\right| << \sum_{i=1}^{m} \hat{b}_i$$
$$\left|\hat{b}_{\beta+1}\right| >> \left|\hat{b}_\beta\right| \qquad (14.12)$$

(4)   **The unmodelled dynamics test.** With this test the disturbance model can be determined as either white or coloured noise, so the most adequate estimator can be chosen. The interpretation of the prediction error autocorrelation $\Phi_{ee}(\tau)$ is used for this purpose.

14.7.2 *Associated with the estimation process*

In this case, the identification is carried out on-line in real-time conditions, which is the reverse of the general situation for the pre-identification stage.

Indicators

- **Associated with the prediction error:**

(1)   *A priori* error:
$$e(k) = y(k) - \varphi(k)^T \hat{\theta}(k-1) \qquad (14.13)$$

(2)   *A posteriori* error:
$$e(k) = y(k) - \varphi(k)^T \hat{\theta}(k) \qquad (14.14)$$

(3)   Mean error value (hereafter may be referred to both the *a priori* and the *a posteriori* error):

$$\bar{e}(k) = E\{e(k)\}$$
(14.15)

(4)   Error variance:

$$\sigma_e^2(k) = E\{[e(k) - \bar{e}(k)]^2\}$$
(14.16)

(5)   Error variance trend:

$$t_e(k) = \alpha \frac{\sigma_e^2(k) - \sigma_e^2(k-1)}{T} + (1-\alpha)t_e(k-1)$$
(14.17)

(6)   Error variance acceleration:

$$a_e(k) = \alpha \frac{t_e(k) - t_e(k-1)}{T} + (1-\alpha)a_e(k-1)$$
(14.18)

(7)   Autocorrelation function:

$$\Phi_{ee}(\tau) = E\{[e(k)e(k+\tau)]\}$$
(14.19)

(8)   Crossed correlation function u/e:

$$\Phi_{ue}(\tau) = E\{[u(k)e(k+\tau)]\}$$
(14.20)

(9)   Crossed correlation function y/e

$$\Phi_{ye}(\tau) = E\{[y(k)e(k+\tau)]\}$$
(14.21)

(10) Cost function:

$$J(\tau) = e(\tau)^T e(\tau)$$
(14.22)

- **Associated with the V-C matrix**

(1)   V-C matrix:

$$H(k) = [\Psi(k)^T \Psi(k)]$$
(14.23)

(2)   V-C matrix inverse:

$$P(k) = [\Psi(k)^T \Psi(k)]^{-1}$$
(14.24)

(3)   Trace of the V-C matrix inverse:

$$trP(k) = \sum_{i=1}^{2n} p_{ii}$$
(14.25)

(4)   Trace trend:

$$t_p(k) = \alpha \frac{trP(k) - trP(k-1)}{T} + (1-\alpha)t_p(k-1)$$ (14.26)

(5)   Acceleration of the trace.

$$a_p(k) = \alpha \frac{t_p(k) - t_p(k-1)}{T} + (1-\alpha)a_p(k-1)$$ (14.27)

- **Associated with the estimated parameters**

(1)   Estimated parameters average:

$$\overline{\hat{\theta}}_i(k) = E\{\hat{\theta}_i(k)\}$$ (14.28)

(2)   Estimated parameters variance:

$$\sigma_{\hat{\theta}_i} = E\left\{\left[\hat{\theta}_i(k) - \overline{\hat{\theta}}_i(k)\right]^2\right\}$$ (14.29)

(3)   Estimated parameters norm:

$$N_\theta(k) = \sqrt{\sum_{i=1}^{m+n}\left(\hat{\theta}_i(k) - \overline{\hat{\theta}}_i(k)\right)^2}$$ (14.30)

(4)   Norm trend:

$$t_N(k) = \alpha \frac{N_\theta(k) - N_\theta(k-1)}{T} + (1-\alpha)t_N(k-1)$$ (14.31)

(5)   Norm acceleration:

$$a_N(k) = \alpha \frac{t_N(k) - t_N(k-1)}{T} + (1-\alpha)a_N(k-1)$$ (14.32)

14.7.3 *Associated with the controller computation*

The indicators associated with the controller computation in an adaptive loop are mainly oriented towards the detection of incompatibilities between the estimated process poles and zeroes and the controller being designed [5].

Indicators

- **Associated with the model polynomial's poles and zeroes**

(1)   Poles and zeroes: magnitude, arguments, real and imaginary parts.

(2)   Poles and zeroes: magnitude trend.
(3)   Poles and zeroes: magnitude acceleration.

- **Associated with the loop error**

(1)   Loop error:

$$e_b(k) = r(k) - y(k) \qquad (14.33)$$

(2)   Loop error average:

$$\bar{e}_b(k) = E\{e_b(k)\} \qquad (14.34)$$

(3)   Loop error variance:

$$\sigma_{e_b}^2 = E\{[e_b(k) - \bar{e}_b(k)]^2\} \qquad (14.35)$$

(4)   Loop error variance trend:

$$t_{e_b}(k) = \alpha \, \frac{\sigma_{e_b}^2(k) - \sigma_{e_b}^2(k-1)}{T} + (1-\alpha)t_{e_b}(k-1) \qquad (14.36)$$

(5)   Loop error variance acceleration:

$$a_{e_b}(k) = \alpha \, \frac{t_{e_b}(k) - t_{e_b}(k-1)}{T} + (1-\alpha)a_{e_b}(k-1) \qquad (14.37)$$

14.7.4 Associated with the closed loop

In this case the main effort is to detect the loop's stability evolution.

Indicators

- **Associated with the characteristic equation roots**

(1)   Characteristic equation roots: magnitude, arguments, real and imaginary
        parts.
(2)   Characteristic equation roots: magnitude trend.
(3)   Characteristic equation roots: acceleration.

- **Associated with the output signal**

(1)   Mean:

$$\bar{y}(k) = E\{y(k)\} \qquad (14.38)$$

(2)   Variance:

$$\sigma_y^2 = E\left\{\left[y(k) - \bar{y}(k)\right]^2\right\} \tag{14.39}$$

(3)   Output trend:

$$t_y(k) = \alpha \frac{y(k) - y(k-1)}{T} + (1-\alpha)t_y(k-1) \tag{14.40}$$

(4)   Output acceleration:

$$a_y(k) = \alpha \frac{t_y(k) - t_y(k-1)}{T} + (1-\alpha)a_y(k-1) \tag{14.41}$$

(5)   Control signal saturation test. The goal is to determine whether the control signal is saturated during a pre-determined number of sampling periods and whether there is any simultaneous change in sign [5].

Chapter 15

Model-based fault detection: an online supervision concept

P.M. Frank and B. Köppen-Seliger

15.1 Introduction

In this chapter an online supervision concept based on analytical redundancy is introduced. Based on mathematical models of physical systems and measurements taken to control these systems, a fault detection concept is developed which can detect actuator, component and sensor faults. With this algorithm, signals are generated which allow a reliable decision as to whether a fault has occurred or not and in some cases even give information about the nature of the fault. As a practical example the application of this concept to a three tank system is described and results are shown.

15.2 Problem formulation

The safety of technical processes is an issue of increasing importance. More restrictive environment protection laws and, in general, the need to reduce costs require the implementation of modern supervision systems in industrial applications. These supervision systems should allow reliable and early fault detection on the one hand and, on the other hand, they should make suggestions for an effective maintenance schedule. Without such a supporting supervision tool the operator is often unable to analyse the quantity of information coming in and to localise a probable fault. In controlled processes especially, small faults can be suppressed by the controller and thereby do not noticeably influence the signal under control. Such faults are therefore hard to detect in the control room using simple threshold logic.

Supervision or diagnosis concepts based on so-called analytical redundancy are able to achieve the task described in principle. In this context diagnosis first of all means comparison between normal or desired and actual behaviour. From deviations or symptoms probable faults should be detected, located and the cause identified in order to take suitable emergency measures or to adapt the maintenance schedule. The comparison of actual with normal behaviour should in this case go further than simple static balance equations which are often already employed in industry. This is in order to make a diagnosis possible under all operating conditions of the system under supervision.

Instead, dynamic balance equations should be used which simply involves evaluation of the input and output measurements together with a mathematical model of the system. This concept is known as fault diagnosis using analytical redundancy. The mathematical process model describes the process behaviour under normal operating conditions and serves as a basis for the evaluation of the redundancy inherent to the measurements. The measurements employed are those which are already accessible in the control room for control purposes.

The main objective is the supervision of those signals that are already accessible by evaluating them further, not to implement additional hardware in the form of additional sensors. This supervision concept can be implemented purely as software on a digital computer which does not influence the process under supervision at all.

A good overview of the state-of-the-art in model-based fault diagnosis can be found in the literature [3–11].

15.3 Process and fault model

Reliable fault detection requires a process description, in the form of a mathematical model, which is as precise as possible. Such a model can normally be found by evaluation of physical laws such as energy balances. In practical applications it is often very complicated to calculate the exact values of the parameters of the model. In some cases this uncertainty in the parameters can have a similar effect on the measurements as the faults to be detected. Also, measurement noise and other disturbances not representing faults can influence the measurements and thereby make it more difficult to detect the real faults. All these influences are called unknown inputs in the following, in order to distinguish them from faults [3,12].

Faults are basically divided into three different classes depending on where in the process they act on it. Actuator, component and sensor faults are distinguished in Figure 15.1. For fault detection the process model mentioned above will be supplemented by a description of the effect of probable faults. By doing this, assumptions are made as to how the faults act on the dynamics of the process. However, no knowledge is necessary about time behaviour and the size of the faults themselves. One of the tasks of the supervision scheme is to compute this.

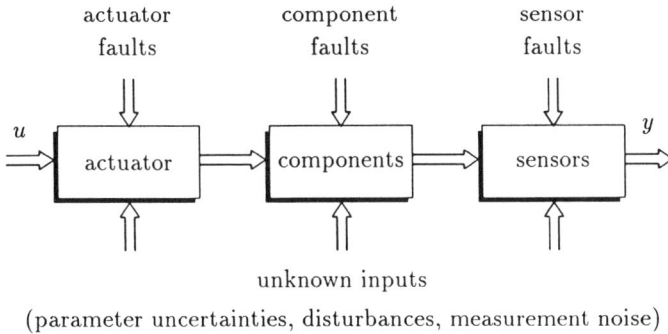

Figure 15.1   Effect of faults and unknown inputs on a technical process

15.4 Robust fault detection

The task of the fault detection scheme is to generate a signal which enables a statement to be made about the appearance of a fault. This signal, the so-called residual, is generated by an observer or filter which computes an estimate of the measured signal corresponding to Figure 15.2. The difference between the measured and the estimated signal yields the residual. The filter is fed by the input and output measurements and contains the process model [12].

The residual should be zero in the fault-free case and non-zero in the case of a fault. Ideally, a comparison of the residual with zero should yield a decision about the appearance of a fault. However, the unknown inputs mentioned above produce a residual which is non-zero even in the fault-free case. Therefore a threshold other than zero has to be employed in order to prevent false alarms.

Another concept to reduce the effect of unknown inputs is robust fault detection. For this, the filter (also called residual generator) is designed in such a way that the faults are decoupled from the unknown inputs, so that the residual is no longer, or is hardly ever, affected by them. Since the residual is then robust against the unknown inputs and only sensitive to the faults, this method is called robust fault detection [12].

This concept of decoupling different effects can also be used for isolating different faults from each other. The filter is designed in such a way that it is sensitive to one fault but insensitive or robust to the other faults. If a filter is designed for each fault in this way, a bank of filters or observers is obtained. Logical evaluation of their residuals leads to a clear decision as to which fault has appeared [2,6,12].

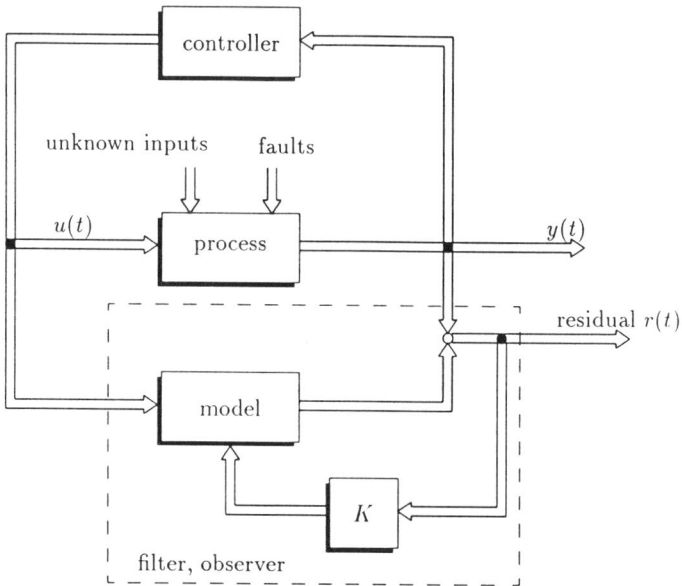

Figure 15.2   General principle of fault detection

15.5 Application

The concept of model-based fault detection is of special interest to the process and chemical industries where aggressive and highly toxic fluids run through tanks, pipes and pumps. Leaks and blockages can have a serious impact on the environment and need to be detected immediately to allow emergency measures to be taken.

The concept presented here has been applied to the laboratory set-up of a three tank system by Prof. P.M. Frank's group at the Institute of Measurement and Control at the University of Duisburg [6,12]. With this method leaks in the different tanks and blockages in the connecting pipes can be detected and localised.

In the following the design of a fault detection filter or observer will be explained for this practical example and results will be given.

15.6 Design of fault detection filters for a three tank system

Figure 15.3 shows the three tank system under consideration [1]. The nonlinear process consists of Plexiglas cylinders, of cross-section A, which are serially interconnected by pipes with stop valves. Tank 2 has a nominal outlet from where water runs into a basin. There it is collected to serve as a supply for the two pumps. The level of the water in all three tanks is measured via piezo-resistive pressure sensors with integrated test amplifiers. With the help of a digital controller, the levels of the water in the left and right hand tanks can be adjusted independently to a given reference level. The level of the third tank always assumes a certain uncontrollable value. For the implementation of blockages and leaks, the interconnecting pipes can be closed by valves whilst additional outlets representing leaks can be opened by valves [1].

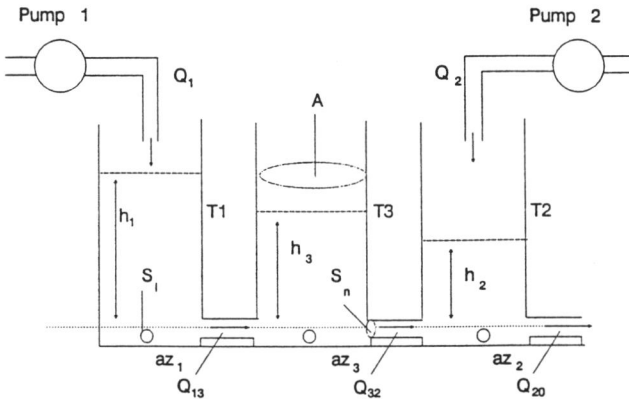

Figure 15.3   Three tank system

For the dynamic model, the incoming mass flows $Q_1(t)$ and $Q_2(t)$ are defined as inputs while the three measurements of the levels $h_1(t)$, $h_2(t)$ and $h_3(t)$ are considered as outputs. The state variables are the water levels of the three tanks such that the full state vector is measurable. The dynamic model is derived via the incoming and outgoing mass flows applying Torricelli's law [12]:

$$A\frac{dh_1(t)}{dt} = Q_1(t) - az_1 S_n \, \text{sgn}(h_1(t) - h_3(t))\sqrt{2g|h_1(t) - h_3(t)|} + Q_{f1}(t) \quad (15.1)$$

$$A\frac{dh_2(t)}{dt} = Q_2(t) - az_3 S_n \, \text{sgn}(h_3(t) - h_2(t))\sqrt{2g|h_3(t) - h_2(t)|}$$
$$- az_2 S_n \sqrt{2gh_2(t)} + Q_{f2}(t) \quad (15.2)$$

$$A\frac{dh_3(t)}{dt} = az_1 S_n \, \text{sgn}(h_1(t) - h_3(t))\sqrt{2g|h_1(t) - h_3(t)|} -$$
$$az_3 S_n \, \text{sgn}(h_3(t) - h_2(t))\sqrt{2g|h_3(t) - h_2(t)|} + Q_{f3}(t) \quad (15.3)$$

$Q_{fi}(t)$, (i=1,2,3) denote the additional mass flows caused by leaks or blockages in the different tanks or the interconnecting pipes. az_1, az_2 and az_3 are the discharge coefficients, S_n is the cross-section of the connecting pipe and g the gravity constant.

The above equations can be rewritten into matrix form with the following definitions [12]:

$$x(t) = \left[x_1(t), x_2(t), x_3(t)\right]^T = \left[h_1(t), h_2(t), h_3(t)\right]^T \tag{15.4}$$

$$u(t) = (1/A)\left[Q_1(t), Q_2(t)\right]^T \tag{15.5}$$

$$B(x(t), u(t)) = \begin{bmatrix} B_1(x(t), u(t)) \\ B_2(x(t), u(t)) \\ B_3(x(t)) \end{bmatrix} \tag{15.6}$$

$$y(t) = \left[y_1(t), y_2(t), y_3(t)\right]^T = x(t) \tag{15.7}$$

$$Q_f(t) = \left[Q_{f_1}(t), Q_{f_2}(t), Q_{f_3}(t)\right]^T \tag{15.8}$$

The components of $B(x(t), u(t))$ are determined by

$$B_1(x(t), u(t)) = (1/A)(-az_1 S_n \, \text{sgn}(x_1(t) - x_3(t))\sqrt{2g|x_1(t) - x_3(t)|} + Q_1(t)) \tag{15.9}$$

$$B_2(x(t), u(t)) = (1/A)(-az_3 S_n \, \text{sgn}(x_3(t) - x_2(t))\sqrt{2g|x_3(t) - x_2(t)|} \\ - az_2 S_n \sqrt{2gx_2(t)} + Q_2(t)) \tag{15.10}$$

$$B_3(x(t)) = (1/A)(az_1 S_n \, \text{sgn}(x_1(t) - x_3(t))\sqrt{2g|x_1(t) - x_3(t)|} \\ - az_3 S_n \, \text{sgn}(x_3(t) - x_2(t))\sqrt{2g|x_3(t) - x_2(t)|}) \tag{15.11}$$

The state-space description results in

$$\dot{x}(t) = B(x(t), u(t)) + Q_f(t) \tag{15.12}$$

and the measurement equation

$$y(t) = x(t) \tag{15.13}$$

Starting from this process and fault models, nonlinear filters for fault detection are designed. The objective is to generate residuals which yield clear decisions even in the case of simultaneous faults. To achieve this a bank of fault detection filters was employed [12]. Three different filters or observers have to be designed, each generating a residual which is sensitive to only one fault and insensitive to the other two. If one neglects the nonlinearities $B(x(t),u(t))$ which are only dependent on the fault-free measurements and the known inputs, then the remaining linear part only consists of three simple integrators excited by the faults.

Considering two of the faults as unknown inputs, the design of a robust filter now becomes particularly simple. The first filter has the following form [12]:

$$\frac{dz_1(t)}{dt} = F_1 z_1(t) + T_1 B(y(t),u(t)) + G_1 y(t) \tag{15.14}$$

$$r_1(t) = L_{1_1} z_1(t) + L_{2_1} y(t) \tag{15.15}$$

with

$$T_1 = (1,0,0); \quad F_1 = -a_{01}; \quad G_1 = (a_{01},0,0); \quad L_{1_1} = -a_{01}; \quad L_{2_1} = (a_{01},0,0) \tag{15.16}$$

The estimation error, as well as the residual, is only a function of the fault $Q_{f1}(t)$ in the first tank

$$\frac{de_1(t)}{dt} = -a_{01} e_1(t) - Q_{f1}(t) \tag{15.17}$$

$$r_1(t) = -a_{01} e_1(t) \tag{15.18}$$

This can also be concluded from the inspection of the transfer function between $Q_{f1}(t)$ and the residual $r_1(t)$. Here we have a first-order lag system with the time constant $1/a_{01}$, which can be assigned freely. With $L_{1_1} = -a_{01}$ the stationary amplification between the fault signal and the residual can be chosen. The faults $Q_{f2}(t)$ and $Q_{f3}(t)$ do not act on the residual, which shows that complete decoupling has been achieved.

A similar design has been made for the second and the third tank [12], yielding:

$$\frac{dz_2(t)}{dt} = F_2 z_2(t) + T_2 B(y(t),u(t)) + G_2 y(t) \tag{15.19}$$

$$r_2(t) = L_{1_2} z_2(t) + L_{2_2} y(t) \tag{15.20}$$

with

$$T_2 = (0,1,0); \quad F_2 = -a_{02}; \quad G_2 = (0,a_{02},0); \quad L_{1_2} = -a_{02}; \quad L_{2_2} = (0,a_{02},0) \quad (15.21)$$

for the second filter and

$$\frac{dz_3(t)}{dt} = F_3 z_3(t) + T_3 B(y(t),u(t)) + G_3 y(t) \quad (15.22)$$

$$r_3(t) = L_{1_3} z_3(t) + L_{2_3} y(t) \quad (15.23)$$

with

$$T_3 = (0,0,1); \quad F_3 = -a_{03}; \quad G_3 = (0,0,a_{03}); \quad L_{1_3} = -a_{03}; \quad L_{2_3} = (0,0,a_{03}) \quad (15.24)$$

for the third filter. The estimation errors and the residuals are again only functions of the fault in the corresponding tank:

$$\frac{de_2(t)}{dt} = -a_{02}e_2(t) - Q_{f2}(t) \quad (15.25)$$

$$r_2(t) = -a_{02}e_2(t) \quad (15.26)$$

for tank 2 and

$$\frac{de_3(t)}{dt} = -a_{03}e_3(t) - Q_{f3}(t) \quad (15.27)$$

$$r_3(t) = -a_{03}e_3(t) \quad (15.28)$$

for tank 3. The observer poles a_{02} and a_{03} are again freely assignable under the condition $0 < a_{0i}$ for a stable observer.

The general structure of the bank of filters is shown in Figure 15.4 [12]. With the help of this bank of filters unique leak detection is possible in all three tanks, although leakages, however, can only be detected uniquely under the assumption of one fault at a time. The fault detection algorithm has been implemented on a PC in the computer language C with a sampling time of 50 ms. The observer poles were chosen to be $a_{01} = a_{02} = a_{03} = 0.25$ corresponding to a time constant of 4 s. Figure 15.5 (a–e) shows experimental results from the online application of the scheme designed to the physical process [12].

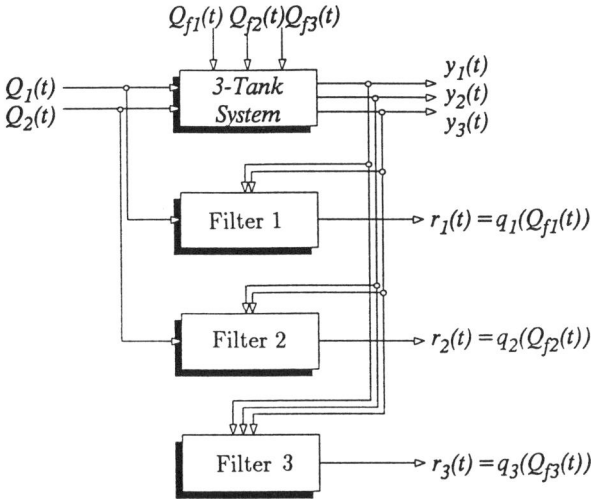

Figure 15.4   Bank of filters for fault detection

The operating conditions can be obtained from Figure 15.5 (a and b), including water levels $h_1(t)$, $h_2(t)$, $h_3(t)$, and incoming mass flows $Q_1(t)$ and $Q_2(t)$. Figure 15.5 (c–e) presents the dynamic behaviour of the three residuals during the following experiments:

- Leak in tank 3 from t=8 s to t=30 s,

- Leak in tank 2 from t=70 s to t=92 s,

- Leak in tank 2 from t=120 s to t=140 s.

The size of the leaks can be obtained from the residuals since they result in a mass flow of 40 ml s^{-1}. The faults can be detected, isolated and even their magnitude determined.

Figure 15.6 (a–e) shows the experimental results for a blockage in the interconnecting pipe between tank 2 and tank 3 for t >3 s [12].

The reaction of the controller can again be seen in Figure 15.6 (b) and the behaviour of the dynamic system in Figure 15.6 (a). Residual $r_1(t)$ shows a slight reaction due to the model's uncertainties. Residuals $r_2(t)$ and $r_3(t)$, however, exhibit a significant response with equivalent magnitude but opposite sign. This means that no two leaks happened at the same time, as the residuals would then have the same sign, but that there is a certain mass flow missing in one tank and too much in the other tank. Therefore a blockage must be the cause of this effect.

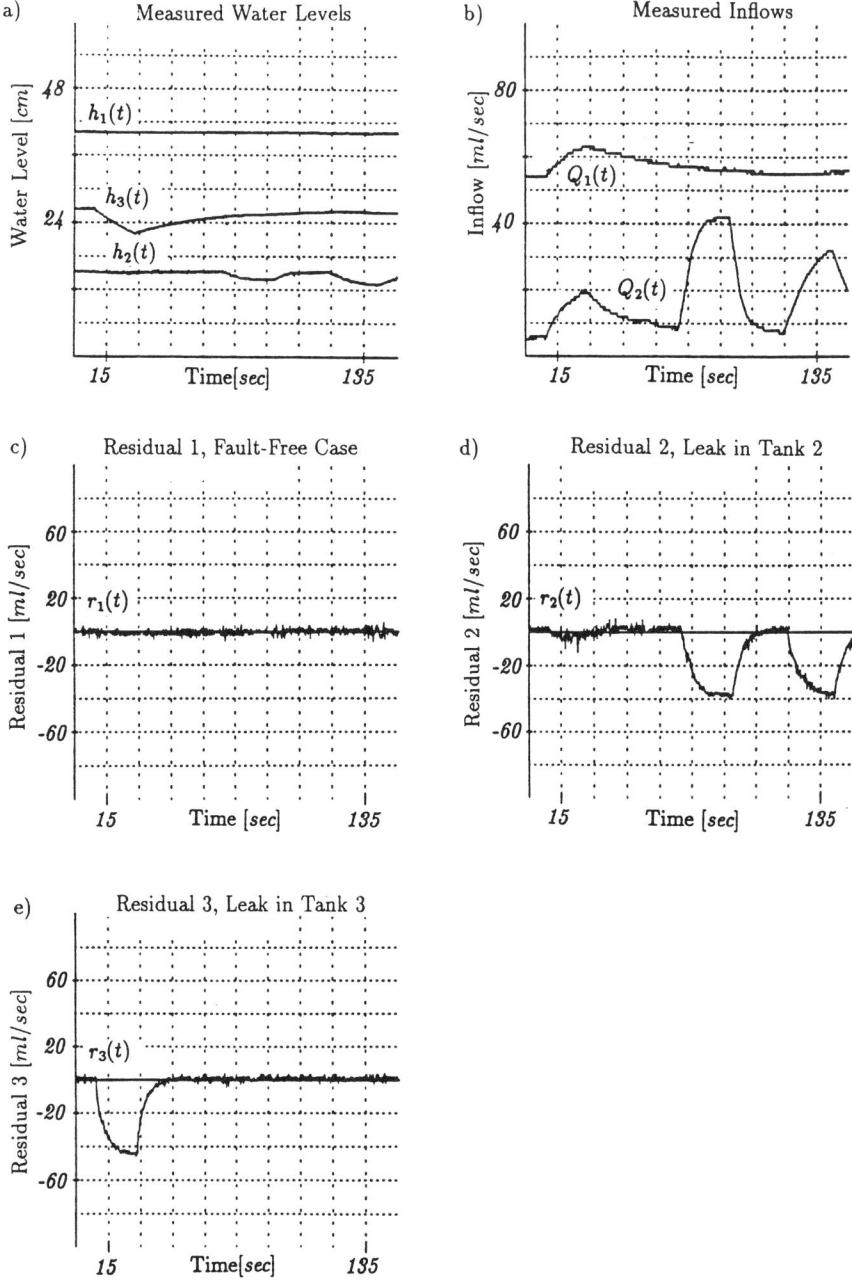

Figure 15.5   Leaks in tank 2 and tank 3

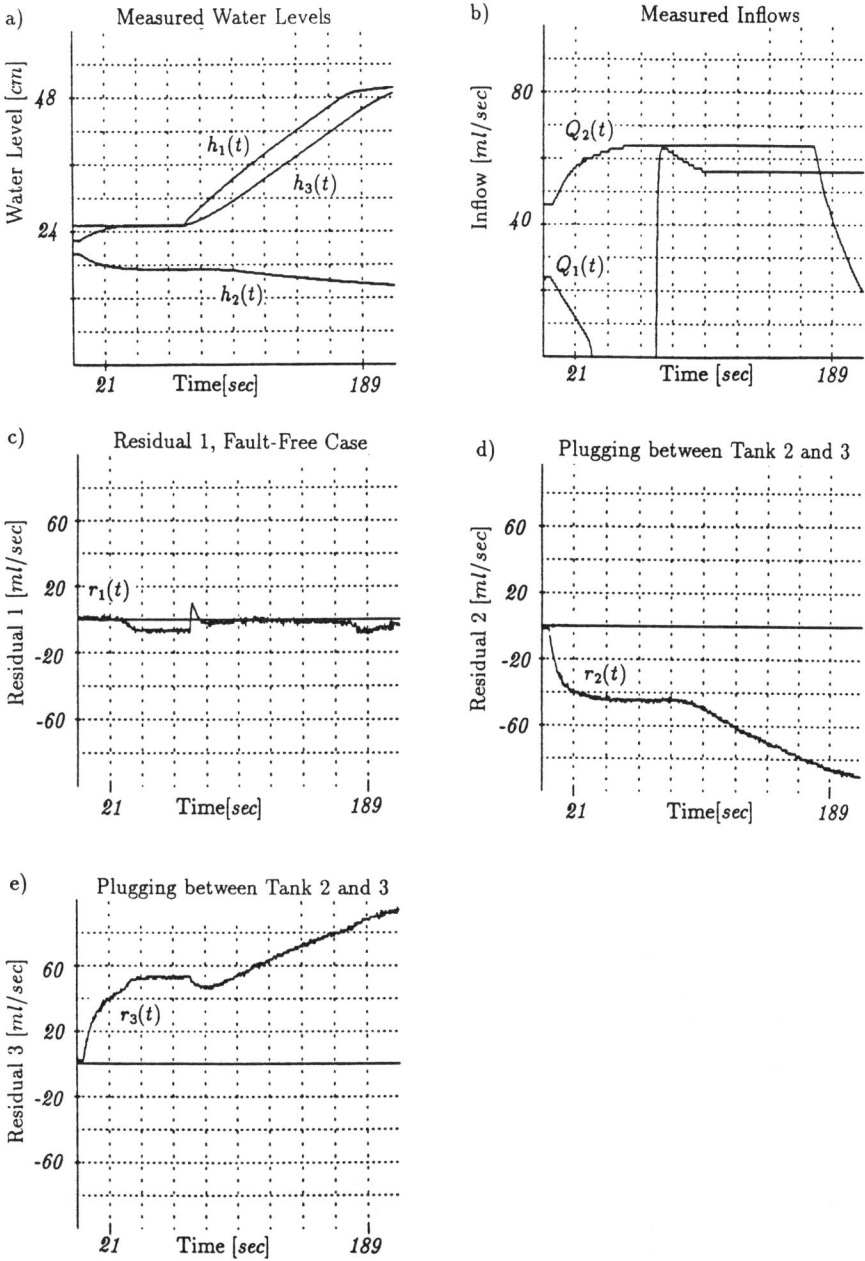

Figure 15.6   Blockage between tank 2 and tank 3

In the case of certain unmodelled faults in the sensors, the residuals react with a significant response. Figure 15.7 (a–e) shows the results for an experiment in the case that the sensor reading for the water level in tank 1 drops to zero for t > 11 s, [12]. An additional logic evaluation of the residuals in Figure 15.7 (c–e) would allow fault identification due to the different time response compared to that of a leak. Since, in that case, the dynamics of the residuals have to be considered, this would need a more sophisticated residual evaluation than simple threshold logic.

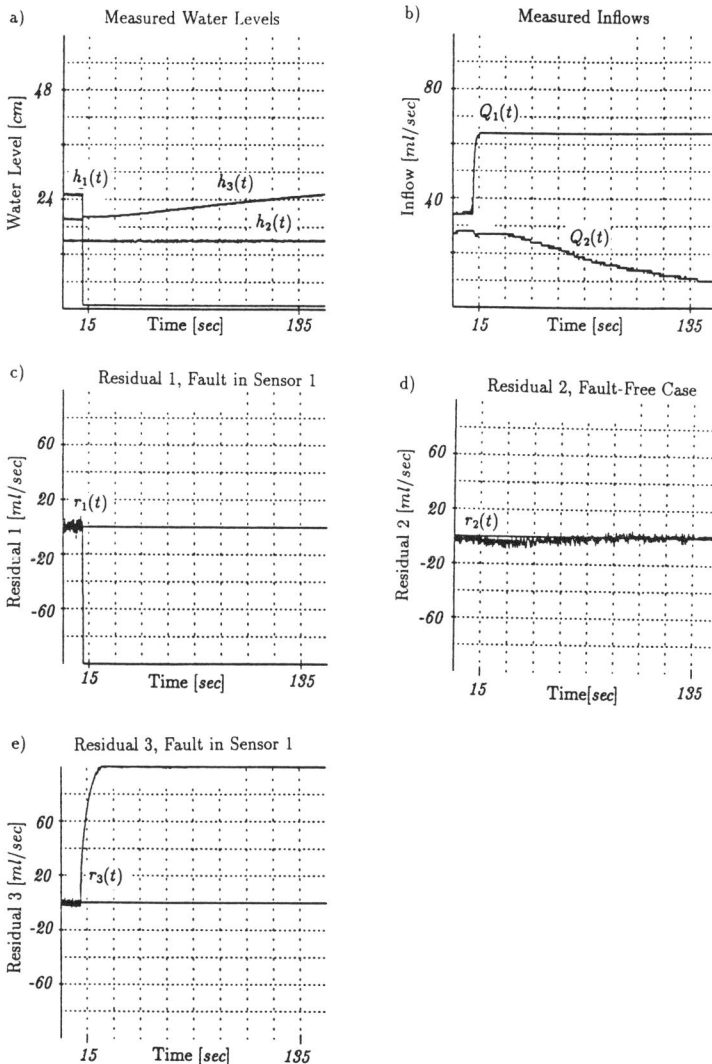

Figure 15.7   Fault in the water level sensor of tank 1

15.7 Conclusions

Due to the simplicity of this example, the design procedure for the fault detection filter is easy to comprehend but it also illustrates its limits. A precise model description, together with the location and number of measurements, is of critical importance. For each technical process a detailed analysis has to show which faults are detectable with which minimal size and whether they can be isolated from each other. The advantages can be seen in the evaluation of information not previously used, namely in the measurements' inherent redundancy. This, together with the process knowledge of the operator, can be used for a new kind of diagnosis without the need for additional hardware.

15.8 References

1   DTS200 Laboratory Set-up Three Tank System, Documentation, amira GmbH, Duisburg, Germany, 1992

2   FRANK, P. M.: 'Advanced fault detection and isolation schemes using nonlinear and robust observers', *Proc. 10th IFAC World Congr.*, München, 1987, **3**, pp. 63-68

3   FRANK, P. M.: 'Fault detection in dynamic systems using analytical and knowledge-based redundancy ñ A survey and some new results', 1990, *Automatica*, **26**, (3), pp. 450-472

4   FRANK, P. M.: 'Enhancement of robustness in observer-based fault detection', *Proc. SAFEPROCESS'91*, 1991, Baden-Baden, pp. 275-287

5   FRANK, P. M.: 'Robust model-based fault detection in dynamic systems', *IFAC Symp. On-Line Fault Detection and Supervision in the Chemical Process Industries*, 1992 Newark, Del., USA

6   FRANK, P. M.: 'Principles of model-based fault detection', *IFAC/IFIP/IMACS Symp. on Artif. Int. in Real-Time Control*, 1992, Delft, Netherlands

7   FRANK, P. M.: 'Advances in observer-based fault diagnosis', *Proc. Tooldiag'93*, 1993, Toulouse, France, pp. 817-936

8   GERTLER, J.: 'Analytical redundancy methods in fault detection and isolation', *Proc. SAFEPROCESS'91*, 1991, Baden-Baden, pp. 9-21

9   ISERMANN, R.: 'Process fault detection based on modelling and estimation methods – A survey', 1984, *Automatica*, **20**, pp. 387-404

10   PATTON, R. J., FRANK, P. M. and CLARK, R. N. (Eds): 'Fault diagnosis in dynamic systems, Theory and application', (Prentice Hall, 1989)

11   PATTON, R. J. and CHEN, J.: 'Parity space approach to model-based fault diagnosis – A tutorial survey and some new results', *Proc. SAFEPROCESS'91*, 1991, Baden-Baden, pp. 239-255

12   WUNNENBERG, J.: 'Observer-based fault detection in dynamic systems', Dissertation Universität-GH-Duisburg, Fortschrittberichte VDI 8 Nr. 222, VDI-Verlag, Düsseldorf, 1990

Chapter 16

Microcomputer-based implementations for DC motor-drive control

C. Lazãr, E. Poli, F. Schönberger and S. Ifrim

16.1 Introduction

The cascade loop is a standard configuration used in DC motor drives. As an alternative, we can use the parallel loop scheme based on independent control of speed and current. Digital implementations of the two control methods are presented here for comparative purposes. Digital control algorithms have been implemented and tested on an IBM/PC or compatible computer provided with a commercial PCL812 interface card.

16.2 Control problem

We want to control the speed of a DC motor with current limitation using cascade and parallel loop schemes. The problem is to obtain good behaviour (constant speed) as the motor load changes on insertion of a load which has inertia. Control algorithms used in the two structures are of the PI type, the integral component needing an anti-reset wind-up mechanism (see chapter 4). With the digital control structure implemented on an IBM/PC or compatible microcomputer (with one CPU), the choice of sample periods of the two loops depends on the expected response times, the execution time of the regulator routines and the times required to measure the speed and the current.

16.3   Technical approaches to DC motor drive control

16.3.1 Cascade configuration

A typical cascade control structure used in DC motor drives [1] is depicted in Figure 16.1.

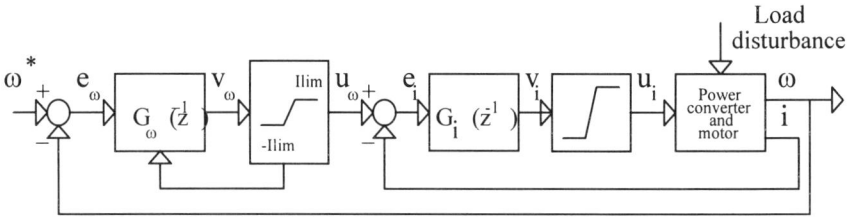

Figure 16.1   Block diagram of the speed control and current limiting circuits for the cascade schemes

The control algorithms used in the current and speed loops, respectively, are of the PI type. They are tuned using the Kessler module criterion [2] and the symmetry criterion [3]. Discretisation of the control algorithms using a trapezoidal approximation [4,5] leads to the following expressions:

$$v_i(n) = v_i(n-1) + \left(\frac{K_{I_i}}{2}T_i + K_{P_i}\right)e_i(n) + \left(\frac{K_{I_i}}{2}T_i + K_{P_i}\right)e_i(n-1) \qquad (16.1)$$

$$v_\omega(n) = v_\omega(n-1) + \left(\frac{K_{I_\omega}}{2}T_\omega + K_{P_\omega}\right)e_\omega(n) + \left(\frac{K_{I_\omega}}{2}T_\omega + K_{P_\omega}\right)e_\omega(n-1) \qquad (16.2)$$

where:

$$e_i(n) = u_\omega(n) - i(n) \qquad (16.3)$$

$$e_\omega(n) = \omega^*(n) - \omega(n) \qquad (16.4)$$

Equation 16.1 is used for the implementation of current controller in an interrupt routine activated with the sample period $T_i = 1.2$ ms. The output v_i of the current loop is limited to the converter limits. The speed loop is activated with the sample period $T_\omega = 5T_i$, with the output calculated from Equation 16.2 followed by a limit non-linearity.

To avoid overshoots due to the saturation of the integral component, the anti-reset wind-up mechanism described in [6] was used. The implementation of a cascade configuration is presented in pseudo-code in Figure 16.2.

```
GLOBAL VARIABLES: COUNT(*), FLAG(**), EXFLAG(***)

INTERRUPT KEYBOARD ROUTINE (*)
{
READ KEYBOARD CHARACTER c;
IF ( c = = ESCAPE ) EXFLAG = 1;
ELSE IF ( c = = CARRIAGE RETURN)  EXFLAG = 2;  STORE c;
   ELSE  /* digits 0, 1, ... , 9, and floating point . */  EXFLAG = 0;
/* acquit interrupt */
}
INTERRUPT A/D ROUTINE
{
APPLY OUTPUT COMMAND
READ CURRENT /* using pacer trigger T₁ = 1.2 ms */
IF (COUNT % 5 = = 0)  {  /* remainder of COUNT/5 is zero, T₂ = 6 ms */
         FLAG = 1;
         READ REFERENCE OF CURRENT LOOP;
         READ SPEED;  /* using software trigger */
         }
CALCULATE CURRENT ERROR;
CALCULATE CURRENT COMMAND  /* relation (1) */
LIMIT THE COMMAND
COUNT = COUNT + 1;
/* acquit interrupt */
}
MAIN PROGRAM
{
INITIALIZE DATA INPUTS, SET CARD TIMER FOR PACER TRIGGER;
INSTALL INTERRUPT SERVICE ROUTINES;
WHILE (1)  { /* infinite loop */
         IF ( FLAG = = 1 )  {
                 CALCULATE SPEED ERROR;
                 CALCULATE SPEED COMMAND;  /* relation (2) */
                 LIMIT COMMAND AS REFERENCE OF CURRENT LOOP;
                 FLAG = 0; }
         STORE MEASUREMENT SPEED & CURRENT IN ARRAYS
         IF ( EXFLAG = = 1 )  GOTO ET;
         IF ( EXFLAG = = 2 )  UPDATE SPEED REFERENCE;
                         /* using characters read from the keyboard */
}
ET:
UNINSTALL ISR'S
DISPLAY SPEED & CURRENT ARRAYS
}
(*)     COUNT is a counter in the current loop, incremented at every 1,2 ms;
(**)    FLAG = 1 means the beginning of the speed loop;
   FLAG = 0 exit from speed loop.
(***)   EXFLAG = 1 stop the process;
   EXFLAG = 2 change speed reference;
   EXFLAG = 0, otherwise.
```

Figure 16.2   Simplified flow diagram for the cascade scheme regulator algorithm

16.3.2 *Parallel configuration*

The cascade system requires that both control algorithms are executed at all times. The parallel scheme depicted in Figure 16.3 uses independent speed and current loops.

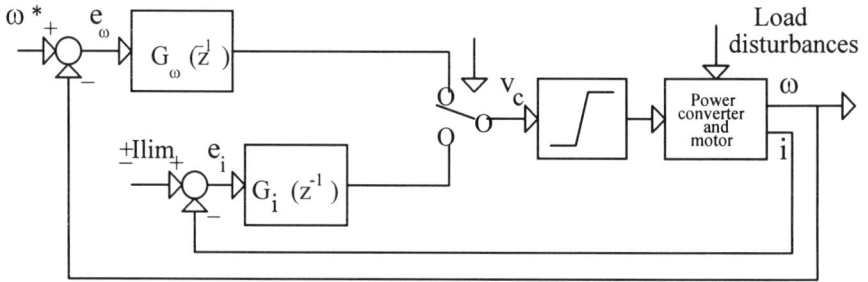

Figure 16.3   Block diagram of the speed control and current-limiting circuits for the parallel scheme

The main difficulty in using parallel control is to obtain the right command when switching from one loop to the other, so as not to cause overshoot [7]. To achieve this, Equations 16.1 and 16.2 are rewritten as:

$$v_c(n) \; = \; v_c(n-1) \; + \; \Delta v_c(n) \qquad\qquad (16.5)$$

where $v_c(n–1)$ is the last converter command value and $\Delta v_c(n)$ is the command increment. In order to determine which loop must be activated, their increments should be compared. If the speed command increment is greater than the current command increment, the current loop is activated, otherwise the speed loop is activated. To implement these algorithms in all cases, the current command must be calculated even in speed mode.

Another test, which verifies whether the current has surpassed the limit, is also performed as protection. The design methods used for current of speed loops are based on Kessler's optimal criteria.

The implementation of the parallel control scheme is presented in pseudo-code in Figure 16.4.

GLOBAL VARIABLES: COUNT(*), EXFLAG(*), MODE(**)
INTERRUPT KEYBOARD ROUTINE (*)
INTERRUPT A/D BOARD ROUTINE
{
APPLY OUTPUT COMMAND
READ CURRENT /* using pacer trigger T1 = 1.2 ms */
IF (COUNT % 5 = = 0) } /* remainder of COUNT/5 is zero, T2 = 6 ms */
        READ SPEED; /* using software trigger */
        CALCULATE SPEED ERROR;
        CALCULATE SPEED COMMAND INCREMENT SCI;
        STORE MEASURED SPEED & CURRENT IN ARRAYS;
        }
IF (SPEED ERROR >= 0)
                CALCULATE CURRENT ERROR WITH +ILIM AS REFERENCE
ELSE   CALCULATE CURRENT ERROR WITH -ILIM AS REFERENCE
CALCULATE CURRENT COMMAND , INCREMENT CCI;
IF (SPEED MODE) { IF (NEW COMMAND EXCEED LIMITS)
                                        MODE = CURRENT; }
ELSE { /* CURRENT MODE */
        IF(| SCI | < | CCI |)   MODE = SPEED; }
IF (CURRENT EXCEEDS LIMITS) MODE = CURRENT;
IF (SPEED MODE)   COMMAND = COMMAND + SCI;
ELSE   COMMAND = COMMAND + CCI;
LIMIT THE COMMAND;
COUNT = COUNT + 1;
/* acquit interrupt */
}

MAIN PROGRAM
{
INITIALIZE DATA INPUTS, SET CARD TIMER FOR PACER TRIGGER;
INSTALL INTERRUPT SERVICE ROUTINES;
WHILE (1) { /* infinite loop */
        IF (EXFLAG = = 1) GOTO ET;
        IF (EXFLAG = = 2) UPDATE SPEED REFERENCE;
}
ET:
UNINSTALL ISR' S
DISPLAY SPEED & CURRENT ARRAYS
}

(*)   The same as in the cascade scheme algorithm;
(**)   MODE = CURRENT means the current controller was chosen,
    MODE = SPEED means the speed current controller was chosen;

*Figure 16.4   Simplified flow diagram for the parallel scheme regulator
                algorithm*

16.4 Discussion

The controllers for both systems were designed to obtain rapid responses without overshoot. The current loop is identical for both systems. As the implementation of both digital control structures uses very short sample periods, it is necessary to analyse the microprocessor execution time efficiency.

From this point of view, the parallel scheme has the advantage of a lower processor utilisation factor in the normal operation mode, the speed mode. Thus, this type of control is more suitable for systems which need to perform additional tasks such as monitoring or adaptive control.

16.5 Laboratory set-up

An electric drive system is represented in Figure 16.5 [8]. The field excitation is fixed and the speed is varied by changing the armature voltage. Two identical three-phase, three-pulse thyristor converters, connected in reverse parallel, are linked between the armature and a three-phase source. The field is separately excited by a single-phase bridge rectifier. External inductors L_1 and L_2 ensure a relatively smooth armature current.

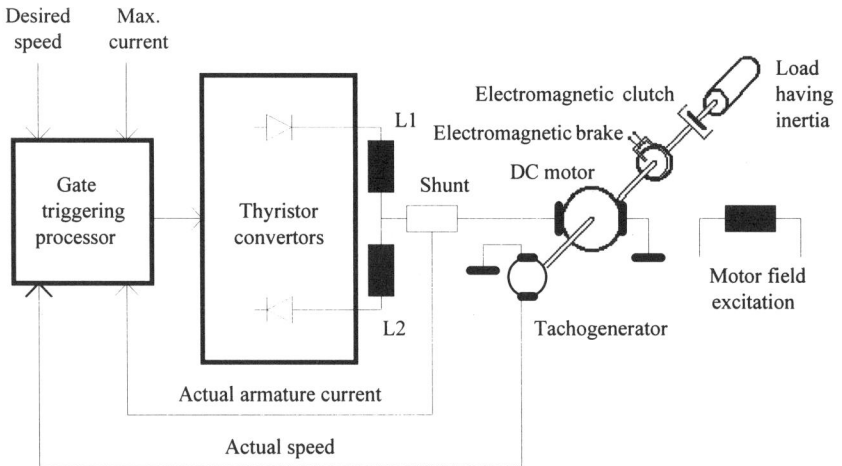

Figure 16.5   Electric drive system for a DC motor

A gate-triggering processor, implemented on an IBM/PC or compatible computer provided with 12 bit A/D and D/A converters, receives the external inputs, actual speed and actual current by means of suitable transducers. In addition, the processor can be set for any desired motor speed and limited to the maximum value of the current.

A DC motor with the characteristics 1.5 kW, 180 V, 11.4 A and 2500 r.p.m. is used. The motor load can be varied from zero to nominal value by means of an electromagnetic brake. Coupling the load with inertia using an electromagnetic clutch, the inertia moment can be modified, and thus the electromechanical time constant of the motor changed.

16.6 Suggested experiments

To test the performance obtained with the cascade and parallel control configurations implemented on the microcomputer, experiments under different operating conditions are carried out. The controllers for both systems are tuned using Kessler's optimal criteria in order to assure low disturbance rejection. We also obtain a very small overshoot and good, fast response. The experiments suggested are to verify the performances of the current and speed loops with speed reference steps and low disturbances.

16.6.1 Responses to a speed reference step

Positive and negative speed reference steps are applied to the two control configurations to consider the transient and steady-state behaviour of the current and speed loops. Both the acceleration and braking states can then be observed.

16.6.2 Responses to load disturbances

A load torque is applied by means of an electromagnetic brake (Figure 16.5) to assess the performance of speed loops tuned by the symmetry criterion. Then a load with inertia is applied using an electromagnetic clutch. As the motor electromechanical constant time is modified, we observe a deterioration of speed loop performance due to the de-tuning of the PI controller for the new inertia moment.

16.7 Illustrative results

For the experiments suggested in Section 16.6.1 we obtain the results in Figures 16.6 and 16.7. In Figure 16.6 the speed and current loop responses to speed reference steps are shown for the cascade loop system, and in Figure 16.7 for the parallel loop system. We notice the good performance obtained in transient and steady-states and the current limitation on the acceleration and braking portions.

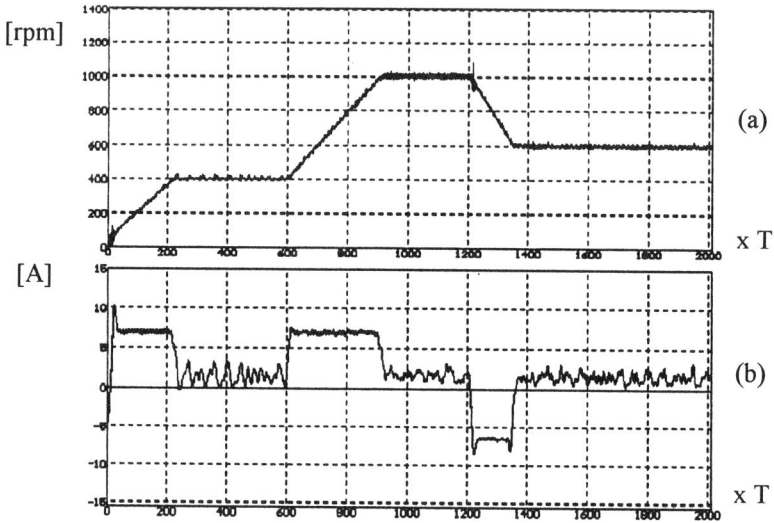

Figure 16.6   Cascade loop system responses to a speed reference step
(a) speed
(b) current

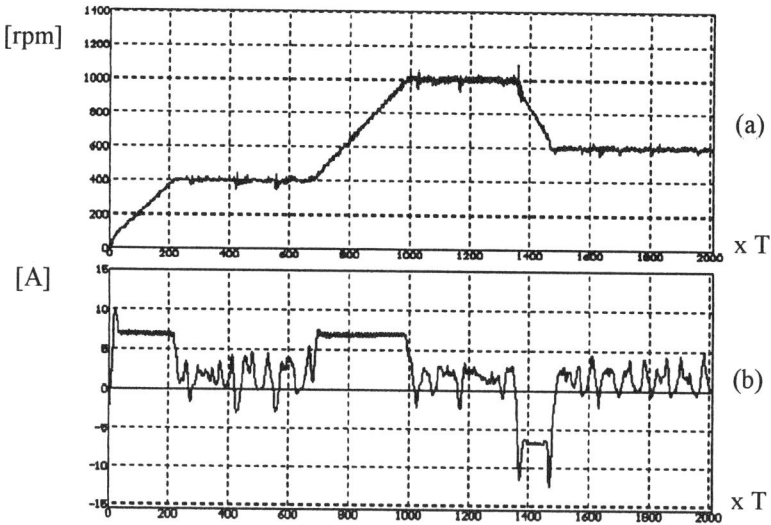

Figure 16.7   Parallel loop system responses to a speed reference step
(a) speed
(b) current

The responses to load disturbances for the two configurations are depicted in Figures 16.8 and 16.9. When a torque load is applied, one can notice very small speed variations for both the cascade control system (Figure 16.8*a*) and the parallel control system (Figure 16.9*a*). The current variations on application of the load disturbance are seen in Figures 16.8*b* and 16.9*b*, respectively. In Figure 16.8*a* and 16.9, we observe the appearance of a larger speed variation when an inertial load is applied.

The oscillations superimposed on the speed and current responses, which are seen in open-loop operation, are due to the motor mechanical system.

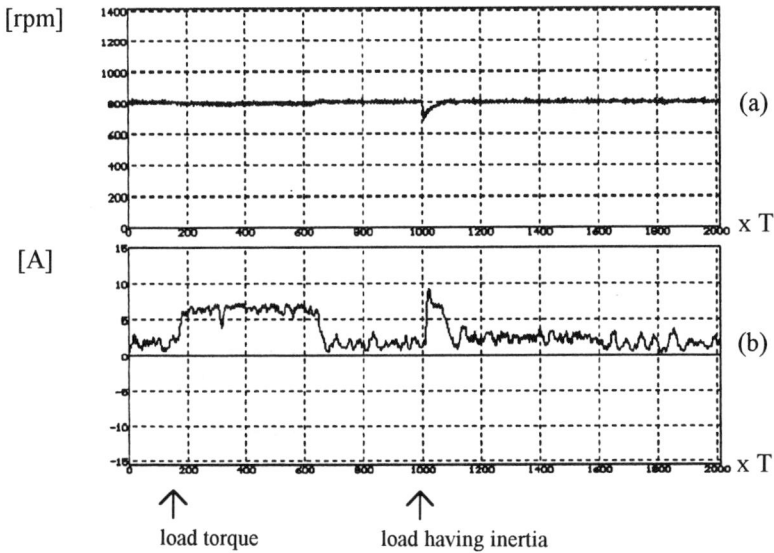

Figure 16.8   *Cascade loop system responses to load disturbances*
(a) speed
(b) current

16.8 Conclusions

Microcomputer-based implementations of cascade and parallel control loops used in DC motor drives have been presented. Experimental results demonstrate that both methods assure speed control and accurate current limiting. They also have comparable dynamic performances to step changes in reference speed. The tuning of speed control loops by the symmetry criterion permits the rejection of load torque-type disturbances. When speed loop performance deteriorates during the application of a load disturbance due to the PI controller becoming de-tuned, other control techniques such as adaptive control could be used. In this case, a parallel scheme (shown in Figure 16.3) is more suitable because it has a lower processor utilisation factor in speed operation mode.

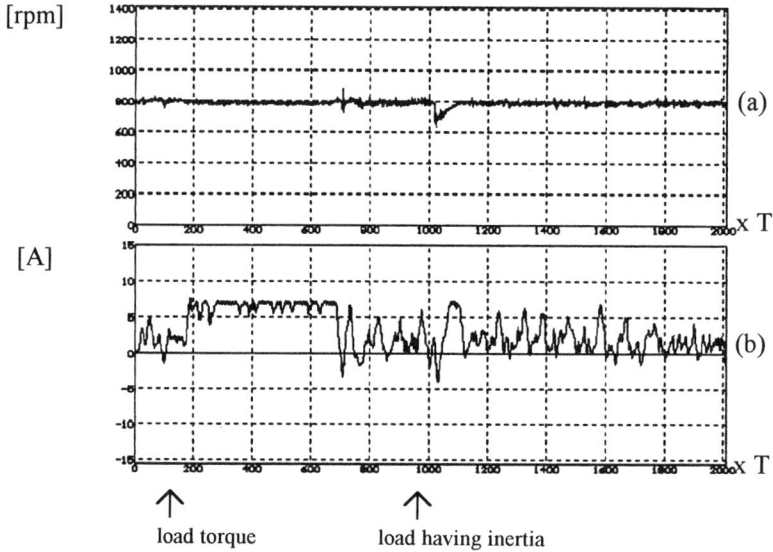

Figure 16.9   Parallel loop system responses to load disturbances
(a) speed
(b) current

16.9   References

1   BIERNSON, G.: 'Principles of feedback control', (John Wiley, 1988)

2   KESSLER, C.: 'Über die Vorausberechung Optimal Abgestimmter RegelKreise', *Regelungstechnik*, 1954, **2**, (12)

3   KESSLER, C.: 'Das symetrische Optimum', *Regelungstechnik*, 1958, **6**, (11)

4   BOLLINGER, J. G. and DUFFIE, N. A.: 'Computer control of machines and processes', (Addison-Wesley, 1989)

5   KATZ, P.: 'Digital control using microprocessors', (Prentice Hall, 1981)

6   ASTRÖM, K. J. and WITTENMARK, B.: 'Computer-controlled systems: theory and design', (Prentice Hall, 1990)

7   JOOS, G., SICARD, P. and GOODMAN, E. D.: 'A comparison of microcomputer-based implementations of cascaded and parallel speed and current loops in DC motor drives', *IEEE Trans. Indust. Appls.*, 1992, **28**, (1), pp. 136–143

8   LAZĂR, C., PĂSTRĂVANU, O. and VOICU, M.: 'Laboratory equipment and process set-ups for control systems', European Seminar on Automation and Control Technology Education, Dresden, 1993

Chapter 17

Software design for real-time systems

A. Braune

17.1 Introduction

The software costs of automation systems tend to increase faster than the hardware costs. Many complex problems are automated with computers using special extensive software products, in which software reliability, flexibility and efficiency play an increasing role. Reusable, clearly arranged and modular software solutions therefore have to be produced, requiring methodological software engineering.

This requirement is considered in the education of university students in the discipline of automation technology, especially by including software design techniques in laboratory experiments in process control lectures. These lectures include some experiments on real-time programming. The objective of this particular experiment is the homogenous, methodological solution of a real-time problem exemplified by a model robot.

17.2 Motivation

The design, implementation and use of software for the solution of control problems can be divided into distinct phases. These phases can be used to describe software characteristics at various points in the lifetime of the software, and the design process should be oriented towards these phases:

— Problem analysis with the requirements analysis
  Analysis tries to answer the question 'What should the software do?'
  Together with the customers, what the input to the system will be, what output the system should produce etc. will be specified.

— System specification with the requirement specification
  The requirement specification is the developer's response to the analysis.
  The functional specification of the visible part of the system is specified by
  the end user.

— Design of the software
  Analysis and specification help to find out what should be done. Design
  specifies how it should be achieved. The result of the design phase is a
  software system that can be implemented with a minimum of difficulties.

— Coding and implementation of the software system
  For most of the history of computing, coding the main program was the
  programmers' main activity. In this phase the programming language is
  selected and the program is coded. If there are several coders the work
  must be divided and co-operation has to be arranged.

— Testing the completed system
  The test phase can involve as much time and effort as the coding phase. A
  program still contains bugs and imperfections. Large systems may not
  even be fully assembled until the test phase begins. The test phase has
  several goals. In the module test, especially, bugs introduced during the
  coding phase are found. The integrating test finds design errors, typically
  involving the interfaces between modules. The final testing step is the
  acceptance test by the end user. Most acceptance problems actually date
  from the analysis phase.

— Maintenance and evolution of the system
  There are several reasons for modifying software during the application
  phase. At first any bugs have to be corrected. A second reason is the desire
  to improve the system and a third category of modifications is adaptive
  changes. Maintenance and modification must be carried out, possibly over
  several years. It can cost two to four times as much as the original coding,
  or up to 80% of the costs in the entire software life-cycle.

Life-cycle models describe the order of these phases. A classical model only
allows us to proceed forwards step by step. An alternative model is based on the
idea that available knowledge is never sufficient to have only one forward
planning of the software. A prototype-oriented model allows some 'trips' through
the entire process of development to obtain experience. Executable test models of
the system, or of particular details, are produced in early phases of the design.

Real-time software includes all the elements of standard software, but it must
satisfy several special requirements in addition. Under real-time conditions several
tasks have to be processed simultaneously and on time. This means real-time
software must be divided into parts or modules with predictable timing constraints
related to task activation, execution and termination. The parallel processes must
be synchronised, if they apply for common resources like the processor or input

and output devices. If parallel processes interchange information, they must be able to communicate with each other.

Therefore the experience and methods of standard software engineering are the basis for real-time software engineering. The useful methods include supplementary possibilities to recognise and to describe parallel processes, synchronisation and communication, timing constraints and control flows between processes.

17.3   Technical approaches

The software development system consists of a real-time operating system and the Top Speed environment with the programming language Modula-2. To maintain homogeneous software technology in all the experiments, a special real-time operating system called XMOD is used. It is a stand-alone system for the Top Speed environment and supports Modula-2 programming. Its programming interface is user-friendly and offers all the important methods for process supervision, interprocess communication, memory management and interrupt and exception handling.

17.4   Discussion

To illustrate a software design, a complex automation problem had to be found which had real-time requirements but which was also practicable in a time-constrained laboratory experiment. The automated process should exist as a real model, not as an analogue or digital simulation. These reasons underlie the choice of using a model robot, in our case, from the *Fischertechnik* company.

The main goal of the experiment is the design and implementation of real-time software, not of special algorithms for robot control.

17.5   Laboratory set-up

The physical model for the experiment, the robot (Figure 17.1), has a special hardware interface for process data. An IBM-compatible PC is connected with the interface and controls the robot. Figure 17.2 shows the main components of the experiment. The robot arm can be moved in three degrees of freedom using three d.c. motors. A smaller motor opens and closes the robot gripper. A positioning system consists of fork-type photo-interrupters. The driving shaft of the gear holds a cup-shard wheel on the circumference of which 32 black and white lines have been printed at regular intervals. When the motor is running, a sequence of pulses will appear at the output of the photo-interrupter which correspond to the dark and bright spots of the wheel and the computer is configured to count every pulse of the photo-interrupter. Each of the three degrees of freedom and the gripper are protected against overrunning the end position by limit keys. The control software must respond to the reaching of a limit key position by switching off the motor.

Figure 17.1   The robot

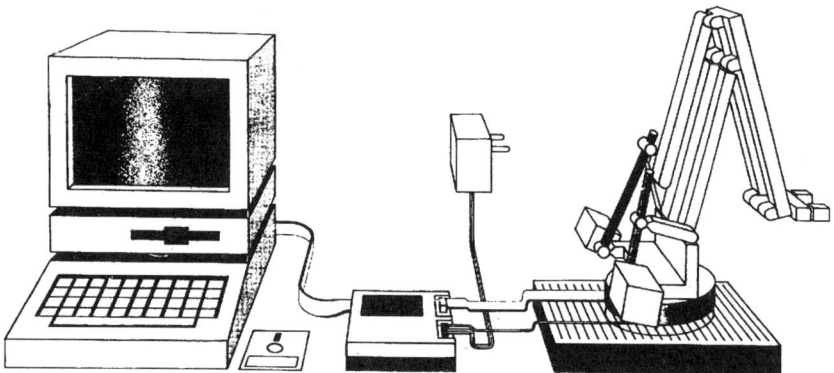

Figure 17.2   Main components of the experiment

The d.c. motors can move the gripper with only one motor speed to the right or left. In this experiment to achieve a variable motor speed the motor is switched on for a constant time (run time) and switched off for a variable time (wait time). The highest speed has a wait time equal to zero. The arm can move with a different speed in each degree of freedom.

The robot interface is connected to the parallel interface (Centronics) of the PC. The computing interface is compatible with the electronic devices of the robot. 8-bit binary input data transfer the information of eight limit keys in polling mode. If the computer scans the input data asynchronously, information will be lost. Three separate binary input signals transfer the information from the photo-interrupter. Each pulse creates a general interrupt and an interrupt process has to find out which motor was moved. 8-bit binary output signals control the motors, each motor has two bits for movement to the right or left. Every data input or output transfers all eight bits. Therefore a data output influences all four motors.

17.6 Suggested experiments

Students have to design, implement and test the real-time control program for the robot. The project is divided into four successive problems:

(1)   To initialise the robot and move the gripper to a home position;
(2)   To move the robot to a fixed position;
(3)   Free robot movement using the keyboard;
(4)   Teach-in programming and recording of the taught steps and motions.

The design aspects are:

— To divide the whole problem into parts, defining modules and module interfaces. Parts of the system analysis and the requirement analysis are given as a description;
— The coding and implementation of the student's own design;
— To test the software with the robot.

The real-time aspects are:

— Timeliness: to count the pulses of the photo-interrupter about every 6 ms and to detect the reaching of limit keys;
— Movement of the robot gripper in the three degrees of freedom, creating three parallel computing processes, necessary because of the variable speed of each degree;
— Synchronisation of the independent movements in the three degrees in one interface process, as the interface needs all control information in one control byte as mentioned above;
— Simultaneity in moving the gripper in all axes while displaying its actual position.

The students' software design follows some aspects of the structured analysis method. This is a data-oriented method and declares (among others) data sources, data drains and marked data flows. The actions for changing data are called knots.

The basis for a design using structured analysis is a data context diagram showing the interface between the software and the real world. Figure 17.3 shows that the robot, the monitor and the keyboard are the data drains and sources. The 'robot control' knot is the action. In the next step this knot is decomposed into a form called a data flow diagram (DFD) which includes more detail. Such a DFD shows the data flows between the knots. Figure 17.4 shows four independent knots without internal data flows. Figure 17.5 represents a part of the refined data flow diagram of knot 1 in Figure 17.4. The structured analysis method demands that the data input and output flows of a knot must be the same in all its refinements. This reduces errors in the refinement steps. Figure 17.6 shows the principle of the method.

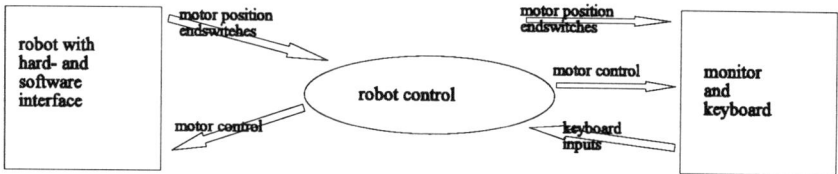

Figure 17.3   Data context diagram
square: data source, data drain
arrow: data flow marked with data names
ellipse: knots (actions) for changing data

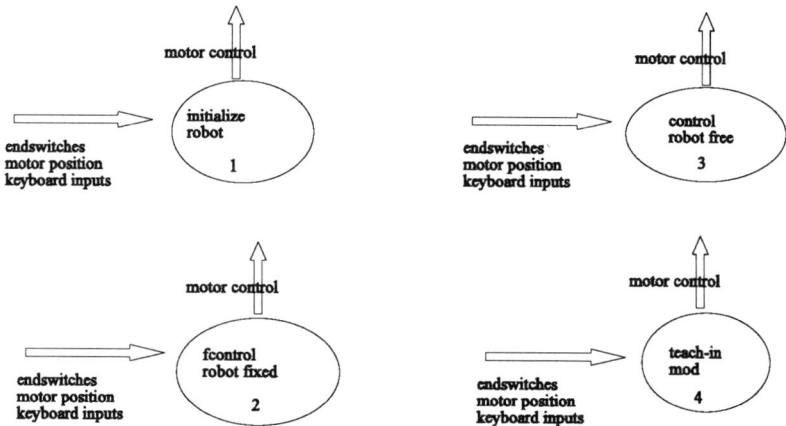

Figure 17.4   Data flow diagram for robot control

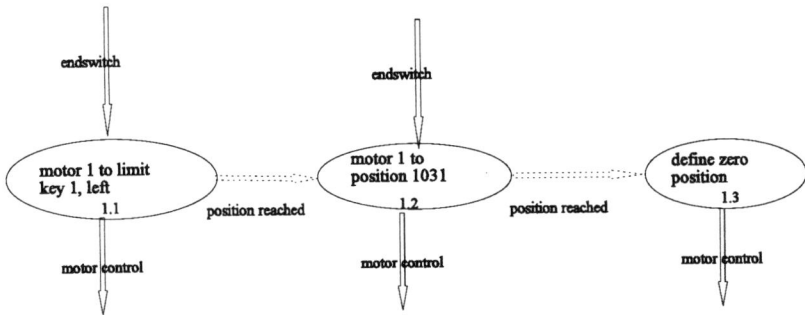

Figure 17.5   Refined data flow diagram
broken arrow: control flow

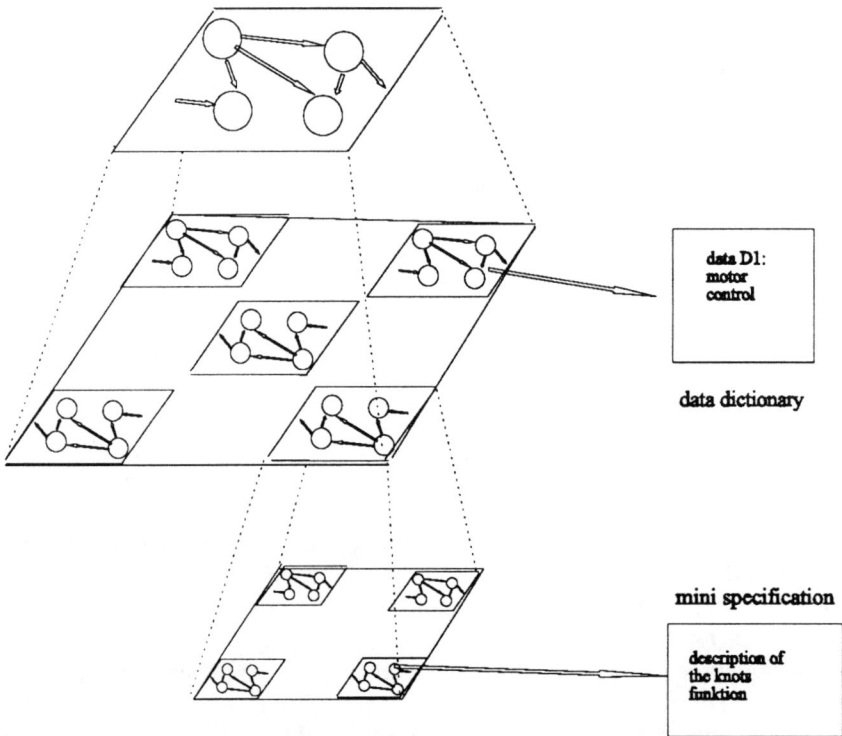

Figure 17.6   Principle of stepwise refinement

Each refinement level has a data dictionary containing descriptions of the newly
defined data in this level. Every knot in the lowest level has a mini-specification

verbally declaring its function. The mini-specification is the basis for coding. A data dictionary for the given software interface is shown in Figure 17.7.

Structured analysis also has a real-time extension, but this is outside the scope of this laboratory experiment.

17.7 Illustrative results

The starting point for a defined robot movement is a zero or a home position. The zero position is marked by the activation of limit keys in all motional axes. Reaching the limit keys, the software sets all internal counters for the motor positions to zero. The gripper will then be moved to a home position, where it points to the side opposite the connecting cable. The arm is angled. The gripper should reach the home position as fast as possible. All motors move without wait times and independent of each other. Therefore each motor has its own motor control process, that implements the control of this motor. Every request for change stores the new motor direction instruction in a mailbox. The process 'prepare motor control byte' finds the changed motor instruction for one axis in the mailbox and combines it with the other unchanged instructions to obtain the final control byte assignment for the interface.

In the second problem the gripper moves from the home position to a fixed final position. The data on the final position are introduced from the keyboard, which is also a data source (like the robot or the monitor), see Figure 17.8. With the keyboard data the control data for axes 1 to 4 are prepared and these actions generate the control bits for an axis, which are later combined to a control byte for all motors. In the real-time design (Figure 17.9) all these actions will be parallel processes. A master process initialises separate slave processes controlling the axes.

In the case of speed control, the gripper should reach the final position on all axes at the same time; a simultaneous speed control is necessary. Movement on whichever axis has the largest distance must be as fast as possible. For the others the 'speed control' task calculates their waiting times so that the whole motor-moving time is the same for all axes. The speed control task stores a new 'wait time' in the 'speed mailbox' and the motor control processes prepare new control assignments for the motors.

The third problem includes free manipulation of the gripper using the keyboard, with or without speed control. This is the preparation for the fourth problem, teach-in programming. Here the gripper learns in sequence a free movement and records the taught movement as often as necessary. Finally the robot should be able to take a workpiece from one position to another, return and grip the next workpiece. Figure 17.10 shows the module design for the experiment and Figure 17.11 shows a coded portion of the main process for robot initialisation.

MPosition: ARRAY [1..3] OF INTEGER
number of counted increments for each motor; each interrupt increments a motor varaible

MCommandMailbox: Mailbox (SIZE(BITSET),4,WAIT)
Mailbox for the single motor control statements

Command: BITSET
Control command for all motors, input for the procedure "Interface Control"

```
          Command:                  ┌7┬6┬5┬4┬3┬2┬1┬0┐
                                     │ │ │ │ │ │ │ │ │
          motor 4  gripper  ─────────┘ │ │ │ │ │ │ │
          motor 4  gripper  ───────────┘ │ │ │ │ │ │
          motor 3  down     ─────────────┘ │ │ │ │ │
          motor 3  up       ───────────────┘ │ │ │ │
          motor 2  down     ─────────────────┘ │ │ │
          motor 2  up       ───────────────────┘ │ │
          motor 1  left     ─────────────────────┘ │
          motor 1  right    ───────────────────────┘
```

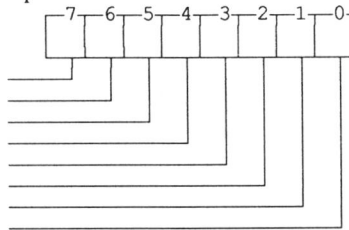

Endswitches: BITSET
Reached limit keys, read from the procedure "rad endswitches"

```
        limit keys :                    ┌7┬6┬5┬4┬3┬2┬1┬0┐
                                        │ │ │ │ │ │ │ │ │
        limit key 8: gripper     ────────┘ │ │ │ │ │ │ │
        limit key 7: gripper     ──────────┘ │ │ │ │ │ │
        limit key 6: motor 3 down ───────────┘ │ │ │ │ │
        limit key 5: motor 3 up   ─────────────┘ │ │ │ │
        limit key 4: motor 2 down ───────────────┘ │ │ │
        limit key 3: motor 2 up   ─────────────────┘ │ │
        limit key 2: motor 1 left ───────────────────┘ │
        limit key 1: motor 1 right ──────────────────── ┘
```

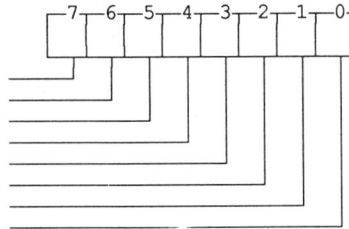

MotorParameterRecord:
RECORD
       motornumber: CARDINAL;
       endswitch : CARDINAL; (final limit key, e.g. limit key 8)
       position : INTEGER; (final position, positiv for the motors 2 and 3, positiv and negativ for the motor 1)
       wait-time : CARDINAL; (for speed control)
END;
All necessary information for moving one motor.

Figure 17.7   Example of a data dictionary

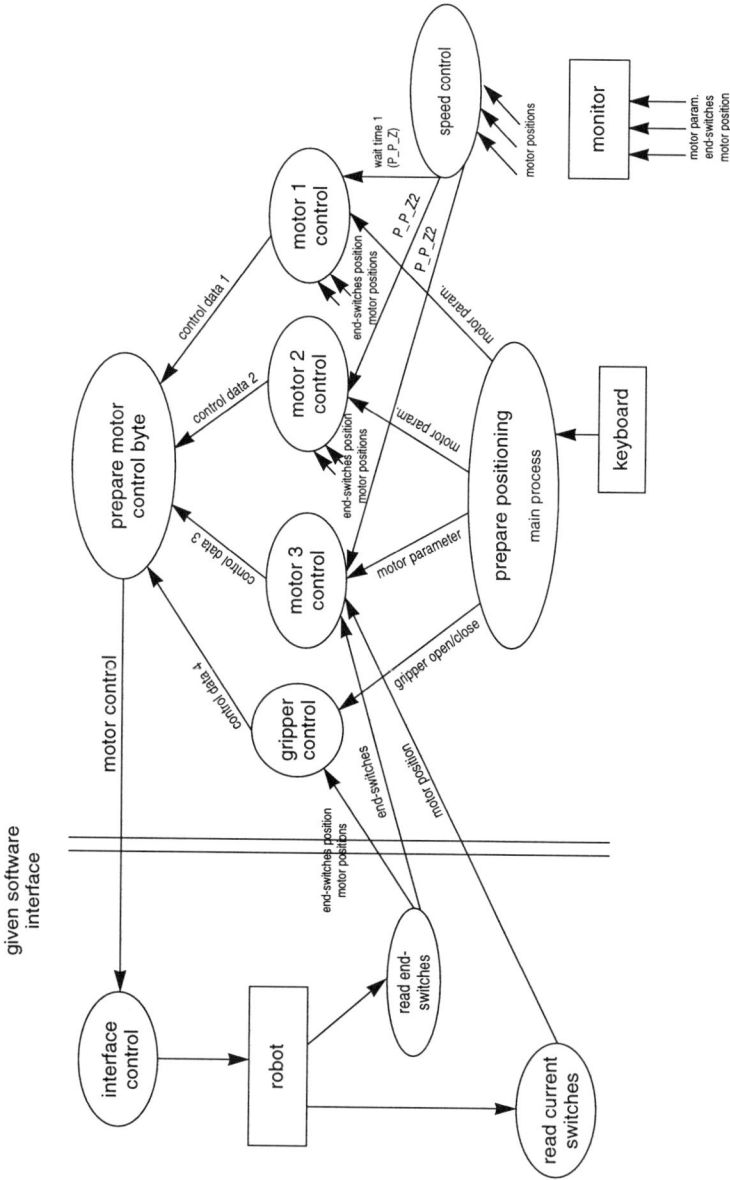

Figure 17.8   Software design for the second problem 'movement to a fixed position'

squares: data sources or drains
ellipses/circles: actions for changing data
arrows: data flows

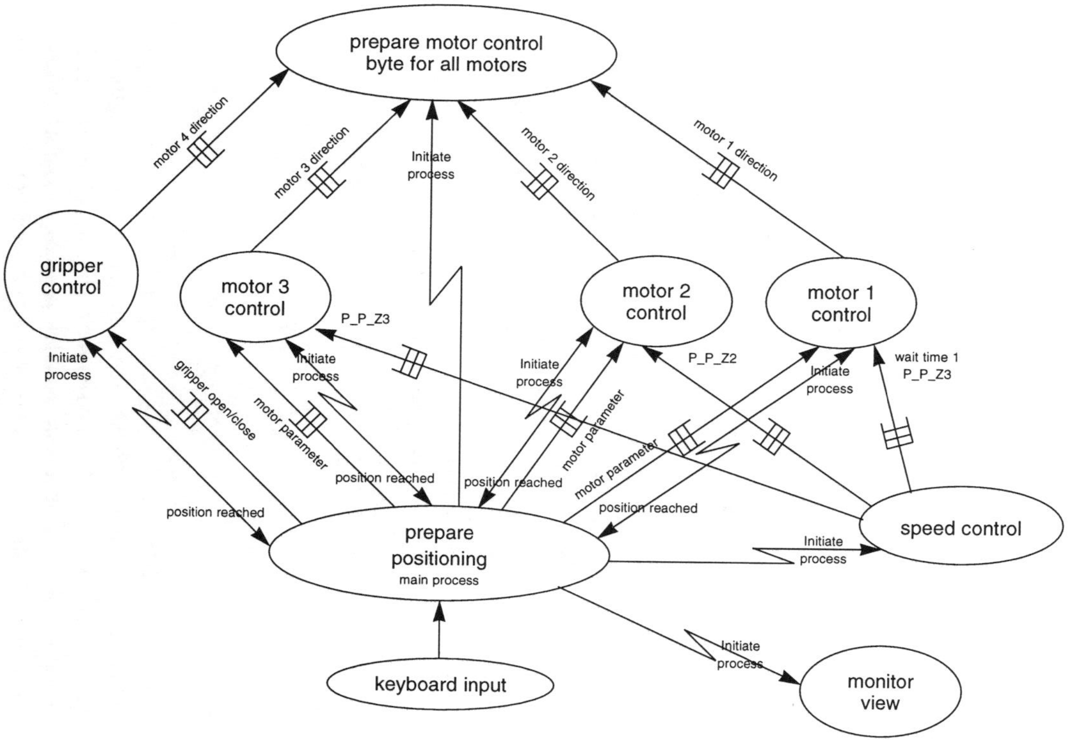

Figure 17.9   Process design for the second problem 'movement of the robot to a fixed position'

ellipse: process

arrow: process synchronisation (initiate or kill a process)

mailbox figure: process communication with the name of the handled data structure

Module Rob_Drv: Hardwaretreiber

GetESwitch	GetLSwitch	SetMotor	EnablePT Interrupt	DisablePT Interrupt

Module Rob_Xdrv: Problemorientierte Treiber

ReadEnd Switch	LSwitch ISR	SendMotor Command	Calc MWZ	Calc GLZ	WriteStatus Task

Programmierschnittstelle

Module Rob_XLib: Steuerung jeweils eines Freiheitsgrades

MotorControlTask					
Motortask					
GotoEnd SwitchTask	GotoPosition Task	StartMotor Prozeß	StopMotor Prozeß	OpenGrab Task	CloseGrab Task

Module Rob_XUPs: Steuerung komplexer (versuchsaufgabenorientierter) Funktionen

RobInit	RobPosition	SpeedContorl Task	Open Grab	Close Grab	WriteStatus Task

Module Rob_TIn : TeachIn Steuerung des Roboters

StartTeachIn	StopTeachIn	Learn Position	LearnClose Grab	LearnOpen Grab

Figure 17.10   Module design for robot control

17.8   Conclusions

Control design for the robot is the final experiment in a real-time programming course in automation. Students design, code and test a complete real-time software system, using elements of 'structured analysis for real-time systems'. They improve their knowledge and skills about parallel processes, about process synchronisation and communication. Evaluating the experience obtained by the first groups of students we can establish:

```
MODUL RobotControl;
(* date, student everybody *)
(* Main modul for the first problem  *)


FROM Xprocess        - IMPORT        NewProcess, Process, KillProcess, NoHandler;
FROM Rob_XLib        IMPORT          GotoEndSwitchTask, GotoPositionTask,
                                     OpenGrabTask, TaskReady;


PROCEDURE RobInit;

VAR     Mproc             : ARRAY[1..6] OF Process;
        Pparam            : TMotorParamRecord;
        PMParam           : PMotorParameterRecord;

BEGIN
(* starting all initializing processes  *)
(*                                       *)

(* motor 1 *)
MParam := FillParamRecord (Mparam, 1, limitkey1, 0, 0, TRUE, TRUE);
(* MotorParameterRecord, motor number, limit key, position, wait-time, reached position=zero
   positiocon , con sider inertia    *)
Mproc[1] := NewProcess (GotoEndSwitchTask, PMParam, 1000, 4000, 3, NoHandler);
(* process code, process parameter, stack size, heap size, priority, exception handler  *)


(*  motor 2   *)
MParam := FillParamRecord (Mparam, 2, limitkey3, 0, 0, TRUE, TRUE);
Mproc[2] :=NewProcess (GotoEndSwitchTask, PMParam, 1000, 4000, 3, NoHandler);

. . .

(* gripper  *)
Mproc[6] := NewProcess (OpenGrabTask,   NIL,  200, 500, 3, NoHandler);


(*  await the end of processes  *)
FOR  i:= 1 TO 6 DO  Request (TaskReady) END;

(* kill all processes *)
FOR i:= 1 TO 6 DO KillProcess (Mproc[i]);

END  RobInit.

. . .
```

*Figure 17.11   Example of code with programming language Modula-2 and
real-time kernal XMOD*

— The necessity for specialised training in software design techniques. Students present and discuss their designed program before coding;
— Development of increased skills in real-time programming using a 'real' robot model. With timing constraints in the millisecond range, many real-time effects occur, e.g. driving through the final position if the actual motor positions cannot be counted accurately, misrepresentation of motor data on the screen, if the representing process does not get sufficient processing time or moves on only one axis, or if the parallel processes receive inappropriate priority conditions.
— High demands on time for students and staff. It will take the students about four to six weeks to do the experiment. Staff inspect the design, the coded program and the program functions.
— Need for good student motivation to do the experiment.

17.9 References

BALZERT, H. (Hrsg.): 'CASE—Systeme und Werkzeuge' (Mannheim, BI-Wissenschaftsverlag, 1990)

LAUBER, R.: 'Prozeßautomatisierung, Bd. 1' (Springer Verlag, Berlin, 1989)

POMBERGER, G., and BLASCHEK, G.: 'Software engineering' (Hanser, Munich)

FIEDLER, J., RIX, K. F., and ZÖLLER, H.: 'Objektorientierte Programmierung in der Automatisierungstechnik' (VDI, 1991)

WARD, P. T., and MELLOR, S. J.: 'Strukturierte Systemanalyse von Echtzeit-systemen'(Hanser/Prentice Hall, Munich/London, 1991)

Jensen & Partners International: 'Top speed version 3.01'. Software Documentation

GOMAA, H.: 'A software design method for real-time systems', *Communications of the ACM*, 1984, **27**, (9)

Index

actuator
   coupled-tank equipment 236
   saturation 61
adaptive control
   basic modules 224
   closed loop 229
   controller design 228, 229
   cost function test 226
   cylinder 179–84, 187–8
   delay/order tests 241
   estimated parameters 229
   evaluation model 224
   experiments 238–40
   expert systems 234
   indicators *see main heading*
   industrial 223
   nonlinear 266–7
   parameter change 229
   programming languages 234
   software 236
   stability test 228
   supervision *see main heading*
   of systems 175–90
   where inappropriate 223–4
aerothermal
   laboratory model 201
   process 203–4
aggregation operators 210
Akaike's final prediction error 11
algorithm
   constant trace estimation 228
   ELS 4
   estimation 2, 4–6, 142

fading 15
fault detection 254
genetic 84, 85–6
LS 4
PD 186
PI 262
PID 6–7
predictive 165–6, 171
recursive identification 15
analytical
   methods 48–9
   redundancy 247
analogue
   computer 31
   controller design 23–41
   simulation 29
anti-wind-up measures 65–6, 263
   compared 68
approximate maximum likelihood 6
artificial intelligence tools 234
augmented system state vector 135
autonomous signal generator
   see under signal

balance equations (dynamic/static)
   248
blowup *see* parameter blowup
Bode diagram 84, 88
boundary control 194, 195, 199–200,
   202–3
bumpless transfer 66
bursting 227